Freeway

Englisch für berufliche Schulen

Ausgabe Technik (Neubearbeitung)

von
Rosemary King
Wolfgang Rosenkranz
Graham Tucker

Ernst Klett Verlag
Stuttgart Düsseldorf Leipzig

Freeway – Ausgabe Technik (Neubearbeitung)

Englisch für berufliche Schulen

von
Rosemary King M.A., Wesel
OStR Wolfgang Rosenkranz, Bad Oeynhausen
Graham Tucker B.A., Castrop-Rauxel

Beratung durch Hellmut Imsel, Stuttgart

1. Auflage 1 5 4 3 2 1 | 2008 07 06 05 04

Alle Drucke dieser Auflage können im Unterricht nebeneinander benutzt werden, sie sind untereinander unverändert. Die letzte Zahl bezeichnet das Jahr dieses Druckes.

Internetadresse: http://www.klett-verlag.de

Redaktion: Volker Wendland, Astrid Keller (Assistenz)
Gestaltung: Thomas Gremmelspacher, Ulrike Wollenberg
Umschlaggestaltung: Nikolaus Keller
Grafik: Christian Dekelver, Weinstadt
Reproduktion: Meyle + Müller Medien-Management, Pforzheim
Druck: H. Stürtz AG, Würzburg
Printed in Germany.
ISBN 3-12-809921-9

9 783128 099217

Freeway

Freeway, das einbändige Lehrwerk für den Englischunterricht an berufsbildenden Schulen führt aufbauend auf der Fachoberschulreife zur Fachhochschulreife. Es entspricht den Richtlinien und Lehrplänen für berufliche Schulen der verschiedenen Bundesländer.

Freeway - Ausgabe Technik (Neubearbeitung) - ist insbesondere an gewerblichen Fachoberschulen und Höheren Berufsfachschulen sowie in entsprechenden Kursen an Fachschulen und Berufsschulen einsetzbar.

Freeway - Ausgabe Technik (Neubearbeitung) - ist wie folgt aufgebaut:

Units 1-5	**Basic structures**
Units 6-10	**Advanced structures**
Units 11-15	**Exam preparation**
Units 16-20	**Exam preparation (Further reading)**

Dem unterschiedlichen Kenntnisstand und den verschiedenen Voraussetzungen bei Schülerinnen und Schülern der angesprochenen Schultypen wurde durch eine flache Strukturprogression in den ersten fünf Units Rechnung getragen. Damit wird auch schwächeren Lernenden der Zugang zu den Themen erleichtert.

In den Units 6-10	werden Aufbaustrukturen in einem berufsbezogenen Kontext vermittelt. Es findet hier bereits eine intensive Vorbereitung auf die Anforderungen der schriftlichen Abschlussprüfungen statt.
In den Units 11-15	werden ausschließlich technische Inhalte behandelt. Texte und Aufgaben entsprechen dem Prüfungsniveau, wobei insbesondere die Beschreibung von technischen Anlagen und Abläufen im Vordergrund steht. Die Grammatikschwerpunkte der vorherigen Units werden hier noch einmal wiederholt und gefestigt.
Die Units 16-20	bieten ein erweitertes Textangebot zum Themenbereich Technik und bereiten ebenfalls auf die Prüfungen vor.

Im Anhang befindet sich eine ausführliche Grammatikübersicht, der vorausgesetzte Grundwortschatz, das kapitelbegleitende Vokabular mit Phonetik, die alphabetische Wortschatzliste sowie eine Liste der unregelmäßigen Verben.

Die Hörverständnistexte der einzelnen Units sind ebenfalls am Ende des Lehrwerks abgedruckt.

Freeway – Ausgabe Technik – zeichnet sich durch folgende Merkmale aus:

■ Übersichtlichkeit und schnelle Orientierung durch das Doppelseitenprinzip.
Die Units 1 bis 10 bestehen aus:

Starter	**Einführung in das Thema**	Auf der mit Fotos und Zeichnungen illustrierten Motivationsseite findet eine erste Annäherung an das Thema der Unit statt. Bereits vorhandene Kenntnisse werden durch einfache Übungen aktiviert.
Teil A (Doppelseite)	**Textarbeit**	Anhand von Texten, Statistiken und Übersichten wird in das Thema der Unit eingeführt. Die Grammatikschwerpunkte werden kontextualisiert vorgestellt und imitativ bzw. reproduktiv geübt.
Teil B (Doppelseite)	**Grammatik**	Im B-Teil wird der Grammatikschwerpunkt der Unit trainiert. Die Übungen sind inhaltlich an die Unit angelehnt, das Niveau des Wortschatzes wird hier aber bewusst niedrig angesetzt, um die Grammatik flexibel einsetzen zu können. Der B-Teil wird immer durch eine anspruchsvolle Aufgabe abgeschlossen, die mit Advanced gekennzeichnet ist.
Teil C (Doppelseite)	**Textarbeit**	Der C-Teil dient der Vertiefung des Unitthemas, wobei in der Regel ein längerer Text zu bearbeiten ist.
Teil D	**Hörverständnis**	Der D-Teil schließt jede Unit mit Übungen zum Hörverständnis ab. Dazu werden hier neben handlungsorientierten Aufgaben Sprechanlässe zur Verbesserung der Kommunikationsfähigkeit geschaffen.

■ Flexible Handhabung
Um den Vorkenntnissen und Interessen der Lernenden gerecht zu werden und um die Ziele der jeweiligen Schulform optimal zu erreichen, soll der Unterrichtende die Möglichkeit haben, flexibel mit Freeway zu planen.

Dabei sind z. B. folgende Varianten denkbar:

- Bei leistungsstarken Klassen könnte es sich z. B. bei den ersten fünf Units anbieten, nur die C-Teile zu bearbeiten (Vertiefung des Themas).
- Ebenfalls wäre es möglich, jeweils nur den A- und B-Teil (Einführung des Themas und Grammatikschwerpunkt) zu behandeln.
- Soll verstärkt die Grammatik geübt werden, dann ist auch ein Schnelldurchgang möglich, indem lediglich der B-Teil durchgenommen wird.
- Die Hörverständnisübungen und weiteren Aufgaben im D-Teil können natürlich mit allen Varianten kombiniert werden.

■ Klare grammatikalische Strukturierung

Der Schwerpunkt der Grammatik liegt in den ersten Units auf der Wiederholung der wichtigsten **Tenses**, in den folgenden Units in den Bereichen **Conditional, Passive voice, Reported speech, Participle, Gerund und Infinitive.**

Um die Übersichtlichkeit im Lehrbuch zu erhöhen, wurden Übungen zu Nebenstrukturen in das Workbook ausgelagert. Eine benutzerfreundliche Grammatikübersicht mit Erklärungen befindet sich im Anhang.

■ Intensive Prüfungsvorbereitung

Auf der Grundlage eines längeren Textes dient der C-Teil in den Units 6-10 der systematischen Vorbereitung auf die schriftliche Prüfung mit gezielter Hinführung zur Textproduktion (z.B.: Skills: Writing a summary)

Eine Info-Box zum schnellen Nachschlagen gibt eine Übersicht über die wichtigsten Schritte zur Bearbeitung von **Comprehension questions, Summary, Comment, Translation und Description.**

In den Units 11-20 (Exam preparation) liegt der Schwerpunkt der Prüfungsvorbereitung auf technischen Inhalten.

■ Weitere Merkmale

- Die Themen und der Wortschatz haben einen festen Bezug zur Arbeitswelt.
- Ein inhaltlicher und kommunikativer Schwerpunkt wurde auf die Bedeutung und Verwendung der englischen Sprache in einem zusammenwachsenden Europa gelegt.
- Unter Berücksichtigung des handlungsorientierten Ansatzes wurde versucht, die Lernenden anhand unterschiedlicher Situationen an adäquate Kommunikationsformen heranzuführen.
- Fester Bestandteil der ersten 10 Units ist die Schulung des Hörverständnisses unter Verwendung einer Vielzahl von Sprechern mit verschiedenen Akzenten (auch non-native speakers). Ausgewählte Texte des Schülerbuches wurden ebenfalls von verschiedenen Sprechern gelesen und befinden sich ebenfalls auf der CD.

Freeway setzt circa 1350 Wörter als Grundwortschatz voraus. Diese werden systematisch wiederholt und durch maßvolle Neueinführungen auf einen Gesamtwortschatz von circa 2700 Wörtern erweitert.

Das Lehrwerk Freeway - Ausgabe Technik (Neubearbeitung) umfasst das vorliegende **Lehrbuch**, ein **Arbeitsbuch** mit einem technischen Teil zu jeder Unit und weiteren schriftlichen Übungen zu Wortschatz und Grammatik, einen **Lehrerband** mit didaktisch-methodischen Hinweisen und Hintergrundinformationen sowie dem Schlüssel zu allen Aufgaben des Schülerbuches. Eine **CD** mit allen Hörverständnisaufgaben und ausgewählten Lesetexten des Lehrbuches begleitet das Lehrwerk.

Wir wünschen Ihnen viel Spaß und Erfolg bei der Arbeit mit Freeway - Ausgabe Technik (Neubearbeitung).

Autorin/Autoren und Redaktion

General topics

Technical topics

Audio-CD

Listening comprehension

Reading texts

Unit 1

Training for work

LEEDS COMPREHENSIVE SCHOOL

GCSE
General Certificate of Secondary Education

LEEDS COLLEGE

Engineering studies

Social studies

Business studies

BTEC
National Diploma

Public Service

Company

University

a Look at the photos and answer these questions.

1. What do the photos show?
2. What can the students from Leeds Comprehensive School do at Leeds College?
3. What possibilities do they have after college?

b Why are you at college?

1 Come to Leeds College

Leeds College

INFORMATION | PROGRAMME OF STUDY | LINKS | COURSES | NEWS | OPEN DAYS

EMAIL: INFO@LEEDS-COLLEGE.AC.UK
WWW.LEEDS-COLLEGE.AC.UK

Our college is one of the largest in the country. We employ over 400 teachers, both full-time and part-time. Our students are of all ages. About half of them are under 20, but we even have one who is 86 years old. The college runs over 200 courses at all levels. Most of the study programmes take place between nine and five, but we also offer evening and part-time courses.

BTEC National Diploma courses
These full-time courses lead to higher job qualifications. Successful students can also go on to university. Changes are taking place in Europe. On these courses students can learn a lot about Europe and study a European language. This is a great start for a career in a modern Europe!

Costs:
The courses are free for students under 18. Students who live over 3 miles from the college receive a free bus pass. Students over 18 can get a grant to cover part of their costs for books and equipment.
For further details see our homepage, send us an e-mail or write to:

Leeds College
Shaftesbury Road
Leeds LS2 45

Tel. 0532/449536

2-Year-Programme of Study

Engineering Studies (Electrical, Mechanical or Civil)
- Mathematics
- General science
- Electrical/mechanical principles
- Computer studies
- English
- European language (French or German)
- Engineering applications of computer systems (CAD & CAM)
- Industrial control and instrumentation

Business Studies
- Mathematics
- General office skills
- Accountancy
- Computer studies
- English
- European language (French or German)
- Statistics
- European business and finance

Social Studies
- Sociology
- Psychology
- Principles of education
- Health care
- Statistics
- Computer studies
- European language
- English

a Answer the questions below. Start your answers with expressions like:

"In the text/brochure it says ..."
"In the second paragraph you can find out ..."
"In line ... of the first/second ... paragraph you can read ..."

1. How do you know that Leeds College is very big?
2. How old are the students there?
3. When do most of the students have their lessons?
4. Why do the students learn a European language?
5. What can you get if you live far from the college?
6. How can people get in contact with Leeds College?

b Talk about your college. Think about:
size of college / teachers / students / courses / your course / costs

c Make a small brochure or a homepage for your college.

2 Two students meet on the campus

Terry: Hello, Sally. Long time no see. What are you doing here?

Sally: I'm doing a BTEC course.

Terry: And how are you finding it here?

Sally: Right now my big problem is mathematics. There are so many new formulas to learn. And then of course, there's the money.

Terry: I know what you mean. I've got a part-time job with my old firm. I work four hours in the evenings in the packing department. It's hard work, but it helps me pay for my flat.

Sally: Oh, it's easier for me. I still live with my parents. But I've got a weekend job in a restaurant. That gives me a little more pocket money. You know what I mean.

Terry: Which department are you in, by the way?

Sally: I'm in the Electrical Engineering department. First year. And you?

Terry: I'm in the Business Studies department. In my second year. We're studying accountancy, statistics and subjects like that at the moment.

Sally: What foreign language are you learning on your course?

Terry: This year I'm doing German for beginners – and that's my problem. There are so many new words to learn. Guten Tag. Ich … heiße Terry. … möchtest du …

Sally: Wow, look at the time. It's eleven o'clock. I've got a science lesson at five past. I have to rush. See you.

Terry: Bye. … eine Tasse Kaffee?

a Copy the information sheet about the two students and complete it.

	Student 1 (female)	Student 2 (male)
Name?		Terry
Course?		
Department?		
Home?		
Problems?		

b Now write about the two students using the information above.

c Use the expressions in the dialogue and make a new dialogue between these two students.

Mike
- 1st year BTEC course in mechanical engineering
- bedsitter (one-room apartment)
- weekend job as a tennis coach
- problems with French/money

Linda
- 2nd year BTEC course in social studies
- house with three other students
- evening job as a waitress at a local club
- problems with statistics/money

d Now write about your college, home and any work you do.

1 Jill is a student

This is what she says about herself:
"I'm eighteen and I go to Leeds College. I don't have any brothers or sisters. I live in Leeds with my parents. My parents give me £100 a month. I spend most of the money on books and things for college. In the evening I work at Pete's Pizza Palace. I need the money, because I buy quite a lot of clothes. In my free time I don't go out very often.
I usually go to college by bus. Sometimes a friend takes me there in his car. At college we have a lot of subjects. I'm in the second year. I like English and mathematics best, but I don't like French. We don't have German. Very often I do my homework with a friend. Normally we both work hard for our course. It leads to higher qualifications and I hope to find a good job afterwards."

a Now talk about Jill.

Example:
"Jill is eighteen and she goes to Leeds College. She doesn't have any brothers or sisters. She …"

b Do you remember? Ask your partner(s) questions about Jill. Do not look at the text on the left.

Example:
Where – Jill – live?
Where does Jill live?
Jill lives in Leeds.
Who – live – with her parents?
Who lives with her parents?
Jill lives with her parents.

1. Where – Jill – go to college?
2. Who – give – her – £100 a month?
3. What – Jill – do – in the evening?
4. Why – she – work?
5. How – she – get – to college?
6. Which subjects – Jill – like best?
7. Who – do – homework – together with her?
8. How – they – work – for their course?
9. Which subject – they – not – have?
10. What – Jill – hope to find?

2 In the cafeteria

a Guess what the students in the picture are doing.

Example:
Helen's writing.

b What do you think? Ask your friends questions about the students in the cafeteria. What are they doing exactly?

Use: What • How many • Where
What … about • Who … to

Example:
What is Helen writing?
I think she's writing a letter.

3 People at Leeds College

John is a student and Lucy is a secretary.

- usually - normally - on weekdays - from Monday to Friday - every day - often	- do homework - organize meetings - help visitors - listen to teachers - take a lot of calls - do tests
today … right now … at the moment …	

■ Talk about John and Lucy.

1. What do they normally do?
2. What are they doing at the moment?

4 On the phone

■ Fill in the right tense.

Sally: Hi, Tom. Are you busy at the moment?
Tom: Yes, sure. Why?
Sally: What **1. (do)** you?
Tom: I **2. (do)** my homework, of course.
Sally: What again? You never **3. (do)** anything else, Tom. Why you **4. (stay)** home all the time?
Tom: You know that I **5. (prepare)** for my exams and right now I **6. (try)** to do my mathematics. Our teacher **7. (give)** us more than 20 exercises every day.
Sally: And what about me? I **8. (wait)** for your call every evening.
Tom: I'm really sorry about that. Why you **9. (not go out)** with Susan sometimes?
Sally: I **10. (not see)** her very often, because she **11. (study)** for a different exam this year. And Anna **12. (live)** in France at the moment, because she **13. (work)** on her French project. I'm so unhappy …

5 Money problems **Advanced**

■ Fill in the right tense. Use the bubbles for help.

"As you **1. (know)** life **2. (get)** so expensive these days, especially for poor students like us. I always **3. (tell)** Jack – why **4. (go)** you shopping at the corner shop? Things **5. (cost)** much less at the supermarket near the college. It **6. (annoy)** me because I **7. (get)** the feeling that he **8. (not listen)** to what I **9. (say)**. It's terrible. We always **10. (have)** arguments about things like that. Sometimes I **11. (think)** he just **12. (be)** lazy. On the other hand there are other things where Jack **13. (think)** we can save. We **14. (own)** a little car for example. Jack **15. (say)** we **16. (not need)** it. He **17. (be)** quite right. And so we **18. (think)** of selling it. We **19. (have)** to do something. Money **20. (not grow)** on trees."

"Is your daughter having a good time at college?"

"You know, students are always complaining about money but every third student owns a car."

"Nowadays a lot of students are thinking seriously about their future and things are getting more expensive. And as you know, money makes the world go round."

1 Leeds and Dortmund: An exchange programme

a Look at the picture which goes with the newspaper article below. Who do you think the people are?

b Read the article and check if you are right.

c Use information from the text and make a list of four

1. things on the programme in Leeds this week.
2. people the students expect to meet during the week.
3. things the students may worry about before they leave Germany.

City plays host to German guests

by Marigold Prince

"For these young people it's quite an adventure" says head of department Willi Schmitz from the Berufsbildende Schulen in Dortmund, Germany. His group of 16 students are just getting to know their English host partners at a reception at Leeds College.

"For many of them it's their first visit to Britain and their first chance to put their English to the test." Indeed, the group of 18–20 year-old students don't know very much about each other before their first meeting. Usually students in both groups receive a passport photo of their partners and a short introduction form before the exchange. This form gives them information about their partner's leisure interests, plans for the future and their favourite food and music.

"I hope mine's not a smoker who's only into heavy metal!" remarks one young Leeds student to his friend as he passes me with some coffee and biscuits for the visitors.

As well as the new situation with their partners and the language, for one week the group has to cope with the different way of life in an English family. "But," adds Willi Schmitz, who is on his fourth visit to Leeds, "after they get to know each other on the first evening, they usually all get on like a house on fire."

This week their programme includes visits to a number of local firms, trips out to the Yorkshire Moors and the Lake District (including a stop at Borwick Hall on the way) as well as a joint project in the College workshops. "This year both groups are working together on a business studies and technology project on local environment problems," says Peter Stewart, Leeds College's exchange organizer.

"We find that working together on a project is certainly one of the best ways of making really good friends. Many of the students from past exchanges still visit their partners or go on holiday with them and every year there are always one or two students who are especially unhappy when it is time to leave," Peter tells me.

As I leave the staffroom at the College, the German students are collecting their luggage and leaving for the first night with their partners and their families. Already it seems to me that some of their earlier ideas about Britain and the British are changing. I hear a couple of comments from some of the German students. One of them says, if I understand his German correctly, "They seem really friendly." "Yes, and it's not raining, either," replies his friend and smiles. Maybe travel really can broaden the mind!

435 words

d **Which of these statements are true and which are false? Use information from the text.**

1. The first exchange visit to another country is always an exciting experience.
2. Exchange students are well-informed about each other before the visit.
3. One of the Leeds students wants a smoker as a partner.
4. After one meeting, the exchange students generally like each other very much.
5. There are visits to local sights on the week's programme.
6. Students who work together get to know each other well.
7. There are two projects this year, one for the business students and one for the technology students.
8. Students who take part in these exchanges do not often see each other again.

e **Which three of the true statements in exercise "d" represent the main ideas in the text?**

f **In these sentences from the text, who or what do the words in bold type refer to?**

1. … **it's** quite an adventure … (line 1)
2. For many of **them** it's their first visit … (line 7)
3. … **he** passes me with some coffee … (line 25)
4. … who is on **his** fourth visit to Leeds … (line 30)
5. As **I** leave the staffroom … (line 56)
6. … **they** seem really friendly … (lines 63/64)

g **Find in the text the opposites of these words**

guest; work; same; last; arrive; alone

h **Here are some words from the text and their dictionary definitions.**
Find the matching pairs.

1. host	a) something people do together
2. to get to know	b) when things or people change places
3. leisure	c) to make something wider or bigger
4. to cope with	d) person who has guests
5. exchange	e) to meet someone for the first time
6. joint (project)	f) something difficult or dangerous
7. to broaden	g) to manage a situation well
8. adventure	h) free time

i **Now answer these questions.**

1. Why is an exchange visit such an adventure?
2. Why is Willi Schmitz sure that his students can have a good time in Leeds?
3. Why is it important for the students to work together on a joint project?
4. Why do students from the past exchanges still visit each other?
5. Why does one student say "Yes, and it's not raining, either"?

j **Translate lines 56 to 66.**

2 An informal letter

You are taking part in an exchange scheme that your college is organizing.
You already have some information about your partner and now you want to write to him or her to introduce yourself.

Write an informal letter in English.
Begin with 'Dear …' and end with 'Yours …' or 'with love from …'

Include these points:
- yourself and your family
- your interests
- what you like/dislike
- your personality and what you look like
- your town
- the weather in your area

Use these expressions:
I am … /in my family we usually …
I am interested in …
I like/love/can't stand …
Some people say I am …/I look …
There's …/we've got … in our town
it's sometimes …/often … here

1 An exchange visit

a **Listen and then answer these questions.**

1. Who are the speakers?
2. Who is talking about what? Look at the pictures for help.

b **Listen again and correct these statements.**

1. The exchange students are working on the environment project at the end of the week.
2. John thinks that visits to local firms and factories are boring.
3. Klaus thinks you learn more about the British as a tourist.
4. John's mother is not very enthusiastic about the exchange.
5. John believes that the Germans are all serious and hard-working.
6. Klaus finds living in an English family strange.

c **Complete the sentences with the phrases in the box below.**

1. … it's really great that we have the chance to see places like that.
2. …, that's why these exchanges are such a good idea.
3. Oh yes, … with you.
4. … that tourists learn half as much.
5. They don't have enough contact with the locals, …
6. They still think, …, that all the Germans are serious.
7. …, isn't it a bit strange for you, Klaus, living in a foreign family?
8. …, you're all so friendly to me.

I suppose • I think • I believe • I'm sure • in my view • I agree • in my opinion • I'm quite certain • I don't think • for example • for instance • I mean

2 Giving your opinion

a Now try to find more arguments for exchanges.
Use the expressions in the box above for your arguments.

b Comment on the following statement using the expressions in the box above.
"Learning a foreign language at college is a waste of time."

Just the job

Are people always asking you what you want to be when you leave school? Try our special chart to help you discover which is the job for you!

■ Work out the best job for your partner. He/she should answer 'yes' or 'no' to the questions. It's more interesting if your partner closes his/her book. When you reach one of the symbols, look at the list below for a suitable job.

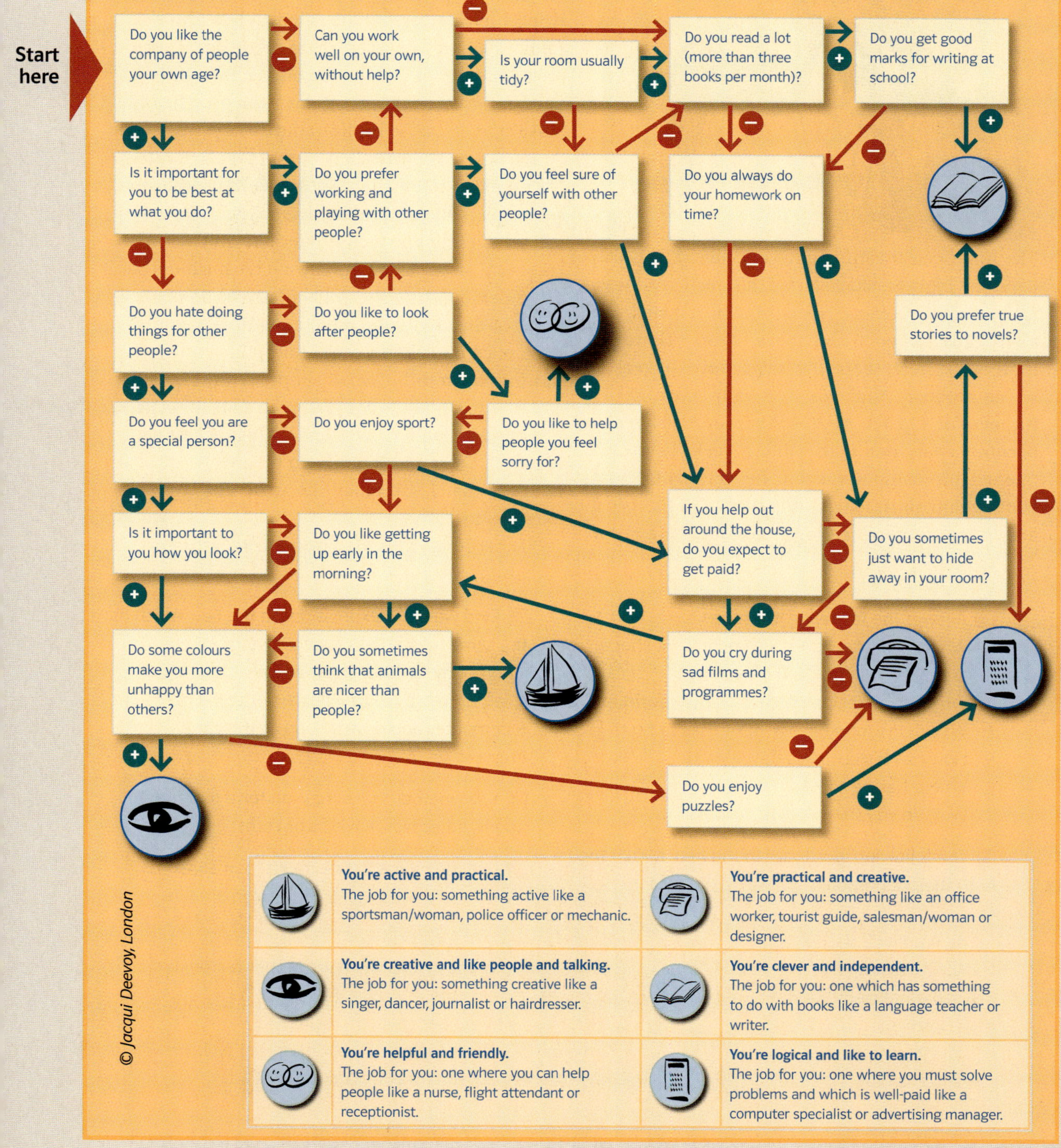

You're active and practical. The job for you: something active like a sportsman/woman, police officer or mechanic.	**You're practical and creative.** The job for you: something like an office worker, tourist guide, salesman/woman or designer.
You're creative and like people and talking. The job for you: something creative like a singer, dancer, journalist or hairdresser.	**You're clever and independent.** The job for you: one which has something to do with books like a language teacher or writer.
You're helpful and friendly. The job for you: one where you can help people like a nurse, flight attendant or receptionist.	**You're logical and like to learn.** The job for you: one where you must solve problems and which is well-paid like a computer specialist or advertising manager.

1 "Computers Unlimited"

a Look at the photo and describe what you can see.

The American dream is alive and living in North London. John Sanson started his own business, Sanson Computers, 6 years ago in a small flat above his parents' butcher's shop. However, in the last few years his sales have increased so enormously that his company now has a turnover of £13.5m. At 28 Mr Sanson has already become a very rich man. Only 2 weeks before he opened his computer shop, he graduated from Nottingham University with a degree in Electrical and Computer Engineering. "A lot of people in my family were self-employed and I wanted to do the same. I have always been interested in computers and while I was at university I saw the possibilities of a growing market in personal computers. So, when I left university, I decided to take the risk," John explains. However, there were not a lot of risks for him; he lived with his parents and was used to life on a student's grant. Moreover, he only had a staff of one, so he didn't have a high wage bill. Sanson's firm is now called 'Computers Unlimited'. He employs more than 80 people who sell and service PC software and hardware. "Nowadays you can go out and buy a computer like a TV or video recorder," says Sanson. As a result his business has expanded enormously. In his first year he made a profit of only £11,000. In the meantime his profits have reached £2.3m a year. Naturally enough, he has moved from his small shop in North London and has now rented a large store in a business park just outside London. He sums up his success story like this: "Of course it has been hard work. But the secret is that I still enjoy the work and my boss likes me, too."

302 words

© Roger Trapp, The Independent, London (adapted)

b Answer the following questions.

1. Where did John Sanson open his first shop?
2. What did John study at university?
3. Why did he decide to start his own business?
4. Why didn't John have a lot of risks at first?
5. How has his business developed?
6. Where has he moved to?
7. Why, do you think, has he moved?

c The following words are in the text. Match them with their definitions.

1. to employ (line 22)	A. the technology of electricity (n.)
2. electrical engineering (line 10/11)	B. to give work to. They ... 150 workers in that factory (vt)
3. self-employed (line 12)	C. to a great extent (adv.)
4. enormously (line 26)	D. (good fortune) to have great ... in life (n.)
5. to increase (line 5)	E. to make or become bigger in size or number (vt)
6. success (line 31)	F. having your own firm or company (adj.)

d Find words or expressions in the text which have the same meaning as the following.

1. company 2. little 3. grown
4. left university with a degree 5. workers
6. computer equipment 7. up to this time
8. employer

e Say whether the words you have found are nouns, adjectives, adverbs or verbs.

f Use the words to talk about John Sanson.

2 A radio interview

"Young people in Southampton" is a popular programme on Radio Solent, the local radio station. Des Jacky is talking to Roberta Crumb.

DJ: Welcome to "Young people in Southampton". Today we have Roberta Crumb with us in the studio. Roberta, you are 24 and you ran a small video production service here in Southampton. Now could you tell our listeners a little about what has happened to you and your business?

RC: Well, it all started when I saw an advert in the local paper. At that time I was employed at a large electronics firm near here.

DJ: And what was the advert for?

RC: "Starting your own business." The local bank offered to give advice to young people who wanted to start up their own business. At first I didn't take it all that seriously. But the more I thought about it, the more interested I became. I talked to my parents and my boyfriend about it. They said I should do what I really wanted to do, which was to start up a small video production service. So, after some time I went to the bank and had a talk with the bank manager. He explained all the advantages and also the risks of being self-employed.

DJ: What happened then?

RC: Well, after a lot of thought I made up my mind to start up the business. We recorded customers' weddings, birthday parties – anything like that, but I didn't earn as much as in my old job.

DJ: And what happened then?

RC: Well, I still had to pay back a lot of money to the bank for the video equipment, and advertising cost me a lot, too. I couldn't live on the income from my work. So I decided to close down the business. I think my main mistake was that I didn't watch the market development carefully enough. People nowadays have their own video cameras and don't need a professional service.

DJ: And where are you working now, Roberta?

RC: I've got a job in a photographer's shop. The pay's not bad, but the problem is I have to pay back the money to the bank. I haven't had a holiday for 2 years now. I just can't afford one.

a True or false?

1. Roberta worked in a photographer's shop.
2. The local bank put an advert in the newspaper.
3. Roberta earned enough money from her business.
4. The bank manager gave her some advice on starting her own business.

b **Complete these sentences. Choose from the nouns and verbs in the boxes. Use the same tenses as in the dialogue.**

1. Roberta … a video production …
2. She … an … in the local newspaper.
3. She … a talk with the …
4. Roberta … up the business …
5. She (not) … the market … closely enough.
6. Roberta (not) … enough customers.
7. She … down her …
8. She … in a … shop.

have (2) • see • run • start • watch • work • close

advert • development • service • business • bank manager • photographer's

c **Join the sentences with the linking words in the box.**

Example:

1. In the interview we hear that Roberta ran a video production service.
2. First she …

In the interview we hear that …		
Then …	After that …	
However …	As a result …	
So …	Now …	First …

1 Softright Systems

Jane Topper moved to Glasgow seven years ago. There she met Harry Coe who didn't like his old job in an import company. As they were both interested in computers they **1. (decide)** to start a small software company. First they only **2. (write)** computer programs for Harry's old import company, but soon they **3. (see)** that a lot of other companies also **4. (need)** special software. So they **5. (hire)** some young programmers who **6. (work)** in small teams and **7. (develop)** a lot of new ideas.

The company **8. (be)** profitable from the start. Although other companies **9. (have)** their problems Softright **10. (not lose)** money. In their third year in business they **11. (make)** a profit of £150,000. Then they **12. (try)** some advertising and it **13. (work)**. That is why they **14. (spend)** more on marketing in the following years. At the beginning Jane and Harry only **15. (have)** two small rooms for their work. After some time they **16. (rent)** offices in several different buildings. But last year they **17. (buy)** a warehouse in central Glasgow. The move **18. (not be)** easy and it **19. (take)** more than two months. "We **20. (lose)** quite a lot of money when we **21. (move)**," Jane said. "But we **22. (not want)** to work in different places any more. That **23. (be)** a good decision!"

a Complete the sentences in the past tense.

b Find the missing parts in the following statements about Jane and Harry's company. Ask a classmate about it. He/she should give you an answer with the help of the information in the box below.

Example:
1. When did Harry's parents move to Glasgow? They moved to Glasgow in 1976.
2. Where …

1. Harry's parents moved to Glasgow in …
2. After college in Glasgow Harry worked at …
3. The company imported …
4. He enjoyed his … very much.
5. But one day Harry left because …
6. He met … on a computer course.
7. She told him …
8. Of course, they needed … for their new company.

- some money
- that she wanted to start her own business, too
- he wanted to be his own boss
- his partner
- 1976
- his father's company
- electronic parts
- work

2 Times have changed

Describe what has happened to the following two companies in the last few years.

Example:
Evans & Sons have been in Sheffield since 2000. Linda Graves has …

	be in Sheffield	sell	make a loss	expand	buy a company abroad
Evans & Sons	since 2000	120 million DVDs so far	never	three times	already
Linda Graves	for five years	about 1,000 cars up to now	once	just	not yet

3 A job interview

Mary Wilson has applied for a job in the export department of Kwik-Fit. In an interview she talks about her life and career.

a Take her role. Use the verbs and the information below.

Example:
"I was born in Leeds in 1979. I went to school from …"

be
pass
get
live
go
study
travel
work

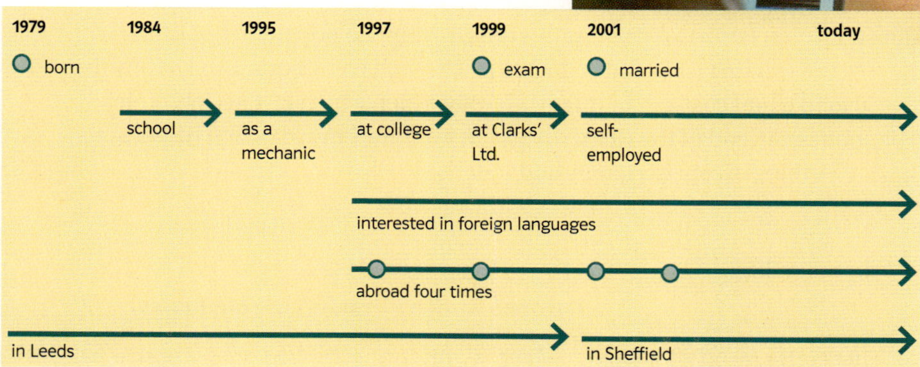

b Ask your classmates about:
date of birth / school / college / important exams / address / trips abroad / experience in English / experience at work / hobbies

c Complete the following report about Kwik-Fit. Put the verbs into the simple past or present perfect tense.

Example:
35 years ago Tom Farmer opened two garages in Edinburgh. In 1974 he …

Kwik-Fit
35 years ago Tom Farmer **(open)** two garages in Edinburgh. In 1974 he **(sell)** to G.A. Robinson. Shortly afterwards Robinson **(get)** into difficulties. Then Farmer **(buy)** the company again. Since then the company's name **(be)** Kwik-Fit. In the beginning most of the business **(be)** with private cars, but in 1987 Kwik-Fit **(begin)** a service for companies. In the last few years customers **(include)** ICI, Avis and British Petroleum. Up to now Kwik-Fit **(keep growing)**.

4 How long …? **Advanced**

a Answer the following questions.

1. How long have you been working with this book? **for … weeks/since …**
2. How long have you known about this college? **for … years**
3. How long have you been learning English? **for … months/since …**
4. How often have you visited England? **once, twice, … times, never**

b Translate the following sentences. Watch out for the correct tenses.

1. Peter arbeitet seit Januar bei Seafood Limited.
2. Er kennt den Manager seit einigen Jahren.
3. Sie planen seit drei Monaten ein neues Projekt.
4. Peters Frau hat sich 3-mal um eine Stelle beworben.
5. Sie wartet schon seit langem auf eine Antwort.
6. Sie sind schon lange nicht mehr zusammen im Urlaub gewesen.

1 Workers – past and present

a In one sentence say what subject these texts have in common.

For the people who lived in the East End of London in the middle of the nineteenth century, life was hard. Often a whole family lived in one room and even 12 year-old children went out to work. Working days were long, often from five in the morning until nine at night. There were no holidays until 1871 when the government introduced the first Bank Holidays. At this time 77 % of the population of Britain belonged to the working class and so had to live in such conditions. Even to buy the most basic food could cost a man three days' pay. When a man lost his job, which could happen at the end of the week, the day or even the hour, or when he became sick, he had nothing except his family and friends to help him.

Europe's workers – underpaid but happy!

This is the result of a survey which a London firm has completed for the managers of a number of multinational companies who have employees all over Europe. The survey shows companies what they can expect from the workers in each country. More than 500,000 workers answered the questions about attitudes to pay, workmates and benefits.

The survey shows that in general Europe's workers enjoy what they do, get on well with their workmates and identify with the company they work for.

However, attitudes to pay are less positive. Although 57 % of workers in the Netherlands are satisfied with their wages, in Switzerland only 44 % of the workers are happy with their pay. In Britain the number is much lower, 35 %. Moreover, the Dutch also feel happiest about their working conditions and benefits. Workers there receive paid sick leave after two days at 70 % of their pay for one year.

The European Union has introduced laws about the maximum number of working hours. However, working hours within Europe still vary quite a lot. In Britain and Luxembourg the maximum is 48 hours, but in Denmark it is much lower, 37 hours.

There are also laws in the EU about the minimum amount of paid holiday a worker can have. The average in Britain is usually 23 days. However, German workers can expect a lot more than this.

Finally, on the question of job satisfaction, the survey's results are positive. Most of Europe's workers are happy with their workmates as well as with the job they do.

The European, London (adapted) **262 words**

Are you happy with your pay?

Are you satisfied with your job?

b Which of these pictures show conditions of work which are mentioned in the texts? Name them.

1
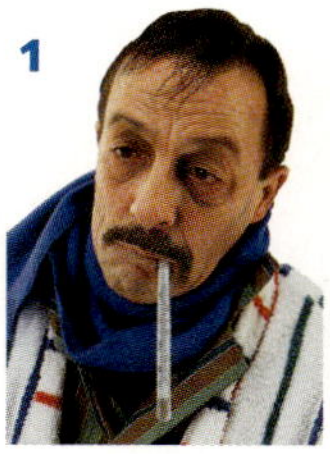
2

3

4

5

6

c The following numbers appear in the texts. What information do they give?
500,000; twelve; 2; thirty-five; 5; twenty-three

d Answer these questions.
1. When did 77% of the British population belong to the working class?
2. How many hours per day did people often work in the past?
3. What help did workers in the past receive when they became ill?
4. Why did the London firm carry out its survey?
5. The survey shows that most workers in Europe agree about some conditions of their work. Which ones?
6. Which two conditions of their work are the Dutch the happiest about?
7. Which conditions of their work do you think the British are unhappy about?
8. Job satisfaction depends on pay. Do you agree? Use the charts to help you with your answer.

e How has the situation of workers in Europe changed in the last one hundred years?

f Translate lines 1 – 8 (Europe's workers – underpaid but happy!).

2 What about you?

a Talk about your experiences.
1. Have you ever worked full-time or in your free time?
2. Have you ever had a job you enjoyed?
3. Is it important for you to enjoy your job?

b Choose five things in this list which you think are important.

workmates	pay	travel	holidays	working hours	to help other people	to be independent	to have a secure job
important							
not important							

In my job	I want to … it's important to me to … I would like to … I hope that …	because in order to … so that …

Job advertisements

A leading European engineering company requires responsible people to work in several departments:

TECHNICAL ASSISTANT
EXPORT ASSISTANT
CHILDCARE ASSISTANT

The successful candidates must be aged 21-30, have a BTEC National Diploma or similar qualifications and some work experience.
A good knowledge of English and German is essential in our international business.
Technical and export assistants should be willing to travel overseas. The childcare assistant should have experience in caring for children under 5.
The company offers interesting jobs with attractive salaries and excellent working conditions.

Applications with full CV to the personnel manager
Mr Len Linscott.

Always one step ahead!

IEP Industries
105-110 Oldham Road
Manchester M13 9PL
Tel: 0161-8342707
www.iepinfo.co.uk

a Answer the following questions.

1. What is the firm looking for?
2. What qualifications must the applicants have?
3. What advantages can you find in the jobs?

b Apply for one of the jobs above. Complete the missing information in the letter below. Use your own personal details, where possible.

name and address
of employer above

your address
date

Dear Sir/Madam/Mr …/Ms …,

With reference to your advert in today's newspaper, I would like to apply for the vacancy of **(job)**. I am **(age)** and of **(...)** nationality. I have attended **(college in ...)** since **(date)**. In **(date)** I hope to pass my examination in **(department)**.
While I was at college/Before college I was employed at **(name of firm/institution)** from **(date)** to **(date)** as a **(job)**. In my job there I **(tasks)**. I am applying for the job as **(name of job)** in your company because **(reasons for applying)**. Enclosed please find my CV and copies of certificates.
I look forward to hearing from you soon.

Yours faithfully/Yours sincerely,

c After you have written the letter of application, the next step is to write a curriculum vitae (CV). Write your own CV including the following information:

personal data, education, qualifications, job experience, interests, references

d Mr Linscott is interviewing two candidates for the job of Technical Assistant. Listen and find out:

1. names of the candidates
2. their favourite subjects at school/college
3. their present jobs
4. why they have applied

e Decide who you would choose for the job. Give reasons.

A survey in Europe: Do you feel European?

A survey released by the European Commission in 2003 shows how Europeans feel about this question. This is what they said:

Source: Standard Eurobarometer 2003

a **Put the letters of these words in the right order to find the names of some European countries.**
canfre; dianler; laity; yemnarg; desnew; umigleb; cegeer; lopdna

b **Answer the following questions.**
1. Which letters on the map show which countries?
2. What is the capital city of each country?
3. What are the most important languages in Europe?
4. Which of the countries on the map are now in the European Union?

c **Do you feel European? Look at the statistics above and make a similar survey for your class. Compare the results.**

d **Find reasons for the differences.**

1 Studying in Europe

Nowadays it's becoming more and more important to spend some time abroad, so if you want to study or train for a job, why not do so in another European country?

"When I arrived in Warsaw they had already organized everything for me," says 20-year-old Sven Berglund from Stockholm in Sweden. "A friend was waiting for me at the station and he took me to the student hostel where I stayed for the first three months. Now I share a beautiful flat with two other students. It's much more comfortable and costs much less than I would have to pay at home. Living in Poland is quite cheap, you know.
At the university in Warsaw, where I'm studying social studies, my sociology lectures are in English and German so I don't have to speak Polish. What I found a little strange in the beginning was that if you want to talk to your professor or practise discussions in seminars you have to wear a suit. I hadn't brought a suit with me so I had to buy one. But normally I wear jeans, of course."

"Don't take everything too seriously," says Brian Auger from Bristol. "That's what I first learned when I started my engineering studies in France. Most of the lectures are in French but I soon got used to that and, of course, it helped a lot that I had learned the language at school for many years.
I spent the first three weeks at a youth hostel here in Lyon because it wasn't easy to find a place to live. Then I got a couple of addresses from a students' organisation and was able to rent a small room. Also I didn't expect such a lot of bureaucracy in France – you have to fill in a lot of forms if you want to live here, and you need several kinds of insurance – and that's not cheap. But if you insure with a students' insurance company you get discounts in fast food restaurants and supermarkets. Isn't that great?"

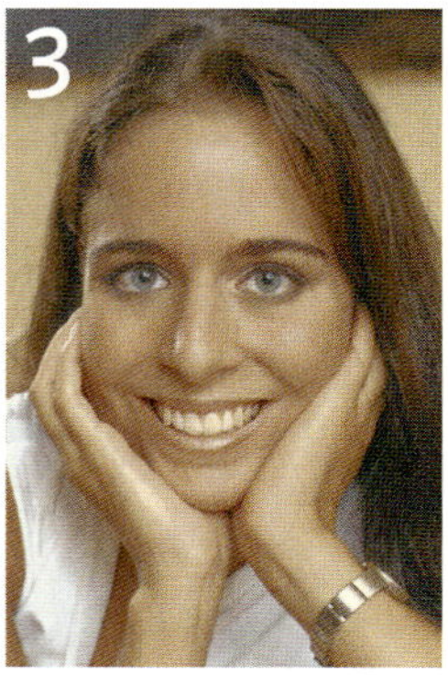

"Studying in Great Britain can be expensive so money is the most important thing if you want to study here," says Esther Rosenkranz from Frankfurt. "First I applied for a place at Oxford University, but I didn't pass the entrance exam. Now I'm at the London School of Economics and I have to pay quite a lot of money a year, but on the other hand you study in small groups right from the start. That means that your professor knows you by name, which wasn't possible in Germany because there are often hundreds of students in the seminars there.
When I applied for a place in business studies I had to take a language test in English first, but I didn't find that too difficult. I've always enjoyed English. I'm really happy here and feel quite at home in the student hostel where I'm living. Luckily it's not too expensive and also it's very international, because quite a lot of the places are reserved for students from abroad, which I find exciting."

a Find at least 3 examples of each of the following in the text:

1. countries
2. cities
3. subjects
4. languages
5. places to live

b Answer the following questions.

1. Why did Sven Berglund move out of the students' hostel in Warsaw?
2. What did Sven Berglund find strange at first and how did he solve the problem?
3. Why isn't it a problem for Brian Auger that his engineering lectures are in French?
4. What disadvantages are there if you study in France?
5. What benefits do you get in France if you insure with a students' insurance company?
6. Why isn't Esther Rosenkranz studying at Oxford University?
7. What differences are there between studying in Germany and Britain?
8. What does Esther Rosenkranz like about the hostel where she is living?

c Find verbs or adjectives in the text which go with the nouns in the box below. Say where you have found them and whether they are verbs or adjectives?

Example:
I have found the word 'train' on line 2. It is a verb.

training • organisation • practice • insurance • application • payment • knowledge

importance • beauty • stranger • possibility • difficulty • excitement

d Use the nouns above to complete these sentences.

1. Sven, Brian and Esther recognised the … of spending time abroad.
2. Sven had no problems with the … of his trip, everything had been done for him.
3. Esther had no … with her English language test.
4. Esther's … for a place at Oxford wasn't successful.
5. In Germany there is very little … that a professor knows his students by name.

e Would you like to live and work or study in another European country? Why? Why not?

2 The European Union

European unity has been a slow and sometimes difficult process. After the 2nd World War most people in Europe dreamed of a United States of Europe: democratic, peaceful and wealthy. However, many governments put their own political and national interests first. These factors have often slowed down the building of European unity. On the right you can read about some of the important steps on the road to a united Europe.

a Complete the sentences in the left column with the information in the right column.

b Now give a report on the history of the EU in the past tense. Use the linking words on page 19 to join the sentences.

c What has happened in the EU since 2003?

1957 Belgium, France, Germany, Italy, Luxembourg and the Netherlands …	… join the European Community.
1958 The European Community …	… sign the Treaty of Rome.
1973 Denmark, Ireland and Great Britain …	… become members of the EU*.
1979 First direct elections for the European Parliament …	… comes into operation.
1981 Greece …	… joins the EU.
1986 Portugal and Spain …	… becomes the single currency of many EU states.
1993 The twelve member states …	… take place.
1995 Sweden, Finland and Austria …	… introduce the Single European Market (The European Union).
2002 The Euro …	… enter the EU.

*"EC" before November 1993.

1 Travelling in Europe

Last summer Robert Page, a computer salesman, travelled around France by train. Here are his travel notes.

JULY

12 Monday
18.00 Paris Gare dû Nord
evening - hotel (Montmartre)

13 Tuesday
morning - Eiffel Tower, Notre Dame
13.15 John Walters / bûsiness lûnch
(Café des Artistes)

14 Wednesday
morning - shopping
13.10 train to Marseilles (Dan and Mike / workmates)

15 Thursday
morning - old port (very hot)
14.00 meeting / Martin Tûrner

16 Friday
11.00 breakfast / Hôtel dû Roi, rest of day on beach
18.30 flight to Manchester

17 Saturday

a Answer these questions.

1. What time did he arrive in Paris?
2. When did he go to the hotel?
3. What sights did he see in Paris?
4. Where did he meet John Walters for lunch?
5. What did he do on Wednesday morning?
6. Who did he meet on the train to Marseilles?
7. When did he visit the old port?
8. What did he do on Friday morning?

b Make complete sentences with the following words:

Example:
after/arrive/Paris/go to/hotel
After he had arrived in Paris he went to the hotel.

1. after/be to/Eiffel Tower/meet/John Walters
2. after/meet John Walters/have/lunch
3. catch/train to Marseilles/after/do/some shopping
4. after/get on/train to Marseilles/meet two workmates
5. after/arrive/in Marseilles/visit/the old port
6. have/a meeting with Martin Turner/after/see/the old port
7. after/have/breakfast/spend/day on beach
8. catch/plane to Manchester/after/spend/day on beach

2 Some business trips do not go as well as others

Andy Williams met Robert Page and told him about what had gone wrong on his trip.

Use past perfect in one clause and simple past in the other. Use the words in brackets to join your sentences.

1. get to platform	train - leave already	(when)
2. the next train - be late	the engine - break down	(because)
3. at the hotel - find	leave my case on bus	(that)
4. get to bus station	it - shut already	(by the time)
5. ask a friend to meet me at hotel	he - not arrive	(although)
6. later - remember	give my friend the wrong address	(that)
7. leave the hotel to go for a meal	lose credit card	(after)
8. return to hotel	someone else - take my room	(by the time)

Example:
arrive at station - ticket office close already (when)
When I arrived at the station the ticket office had already closed.

3 At the airport

Mr Jones, a businessman from Cardiff, went to the airport last week to get his plane to Brussels.

a Make sentences about these situations.

Example:
drive to airport / start to rain
As/While he was driving to the airport it started to rain.

1. show ticket at airport / passport fall out of pocket
2. look around bookshop / meet a friend
3. plastic bag break / he and friend go upstairs
4. friend get coffee / he sit down at table
5. they call his flight to Brussels / he and friend drink coffee

b Jason Norman is a journalist for the "Evening News" in Seattle. Two weeks ago he travelled to Frankfurt. At the airport he went to the bank to change some money.
Put the verbs in these sentences into simple past or past continuous. Be careful, some sentences have both verbs in the simple past. Join your sentences with one of these words: *while, as, when* or *and*.

Example:
A lot of passengers (wait) at the counter - Jason Norman (walk) into the bank
A lot of passengers were waiting at the counter when Jason Norman walked into the bank.

1. he (look) at the exchange rates - man behind him (ask) to borrow a pen
2. he (stand) in line - he (count) his money
3. he (get) to the counter - Jason (hand) the money to the assistant
4. he (talk) to the assistant - there (be) an announcement
5. Jason (listen) to the announcement - the assistant (leave) the counter
6. she (return) - she (give) him a form
7. she (count) his money - Jason (sign) the form
8. she (give) him the money - he (put) it into his pocket

4 On the boat **Advanced**

Put the verbs in brackets into the correct form of the past tense (past simple, past perfect, or past continuous).

While Tim Smith **1. (stand)** and **2. (look)** over the side of the boat which **3. (take)** him to Ostend, he **4. (try)** to imagine the people who **5. (travel)** below him in the Channel Tunnel. Although the tunnel **6. (open)** many years ago and he **7. (have)** enough time to get used to the idea, he still **8. (find)** it difficult to imagine the journey to the Continent by rail and not by boat. Before they **9. (complete)** the tunnel, Britain **10. (always be)** an island and many people still **11. (want)** it that way.

1 Europe's cities lead the way

A long time ago, before people had even thought of ideas like the EU, Europe consisted of a number of independent nations. Nowadays, with more and more countries in the EU, the question is, what developments will tomorrow bring?

For some people, the most exciting aspect is its cities. And why? Well, look at any number of European cities even today and you can see that they are enjoying the greatest growth, energy and activity for a long time.

London for example, is exciting and the most cosmopolitan city in the world – the bridge between Europe and America. Berlin, on the other hand is the link between Western and Eastern Europe, its glass-walled Reichstag a symbol of the new European democracy. And it's not only in Europe's western capitals that you can feel this energy. The former eastern capitals – Prague, Warsaw and Budapest now not only look better after the neglect of the communist era, they also show a desire for the ideals of a modern Europe: tolerance and new ideas. Each city has its own strengths. Talinn in Estonia is rediscovering its strength as one of the great Baltic trading centres, while places like Marseilles, Barcelona and Liverpool are changing their run-down port areas into new centres which will link people to the sea. The diversity of each city is also one of its strengths. Each city is developing its own ideas. In Paris, a mayor once turned one of the main roads into a huge beach with sand and sunchairs instead of cars. The London Eye, also known as the Millennium Wheel, which was put up as a joke for the 2000 millennium, has now become a permanent attraction in London.

Even more exciting are the links between cities. High-speed trains and low-cost airlines are connecting more of these cities with each other. Such links mean that millions of Europeans can afford both the time and the money to visit places they could not have visited before. These links are also bringing Europes's cities closer to each other than to their own countrysides. Londoners will go shopping in Barcelona, while Parisians take the Eurostar to visit the free museums or the clubs in London. City leaders will talk to each other without waiting for the OK from governments.

In future Europe will have to look to its cities for new ideas, new jobs and new leadership. The growing links between cities will help to speed up the disappearance of national boundaries and each city and its people will gradually realize that their interests lie with each other. **431 words**

a Describe what you can see in the photo. It shows something you can find in one of the cities mentioned in the text. Which city do you think it is?

b Which of the following arguments can you find in the text?

1. Europe's strength is its independent nations.
2. The growing number of member states in the EU will increase the importance of Europe's cities.
3. Europe's future lies in its cities because they are places of growth and energy.
4. London is exciting because people from many different countries live there.
5. Berlin is important as a bridge between Europe and America.

6. The neglect of cities during the communist era means that they are not as important as the capitals of western Europe.
7. The strength of each city lies in their common ideas.
8. As a result of better rail and air links, more and more people are able to visit foreign cities.
9. The differences between cities will slowly disappear as a result of links between their people.

c **Which of the arguments you found in the text do you agree with and why?**

d **Answer the following questions on the text in complete sentences.**
1. How has the view of Europe changed over the years?
2. What do London and Berlin have in common?
3. Why are Europe's eastern capitals also exciting?
4. What examples can you find of some new ideas?
5. What strengths do cities have individually and in general?
6. What advantages will new transport links between cities bring?

e **Translate lines 37 to 54.**

f **Project: Write about a European city you know, saying what you like and dislike about it.**

2 How Europeans go on holiday

a **Say which of the two photos below shows what kind of holiday you take and say why.**

Country	Holidaymaking rate (at least a trip of four nights)	Domestic trips	Outbound	Holidaymakers within the EU	Holidaymakers outside the EU
Belgium	40.2%	17.9 %	82.1%	73.9%	26.1%
Germany	76.9%	34.0%	66.0%	66.2%	33.8%
Spain	37.3%	89.7%	10.3%	59.4%	40.6%
Italy	46.3%	77.2%	22.8%	55.3%	44.7%
Netherlands	67.9%	37.0%	63.0%	74.7%	25.3%
Finland	51.9%	71.5%	28.6%	61.0%	39.0%
United Kingdom	60.9%	55.0%	45.0%	65.8%	34.2%

eurostat 2002

b **Find out about the following:**
1. Which holidaymakers travel most/least?
2. Which holidaymakers stay in their home countries most/least?
3. Which nationalities travel within the EU/outside the EU most/least?

c **Answer these questions. What do you think?**
1. Why do some people travel more than others?
2. Why do some people stay in their home countries, why do others prefer to travel abroad?
3. Why do many people stay within the EU, why do others take their holidays outside the EU?
4. Is holidaymaking different in central and southern Europe? Why/why not?

d **Write a text about holidaymaking in Great Britain, Germany and Italy. Use the results from above.**

1 Voting systems

a First listen to these two European politicians who are talking about their own political careers and find out about the following points:
1. Name 2. Country 3. Political party 4. Special interests
5. Constituency 6. Former job 7. Member of Parliament since

b Now describe the two politicians in complete sentences.

c Listen to the first politician's talk again. He went on to talk about the following statistics. What do you think he said?

The election 2001

Labour	40,8 %
Conservative	31,8 %
Liberal Democrats	18,3 %
Others	9,1 %

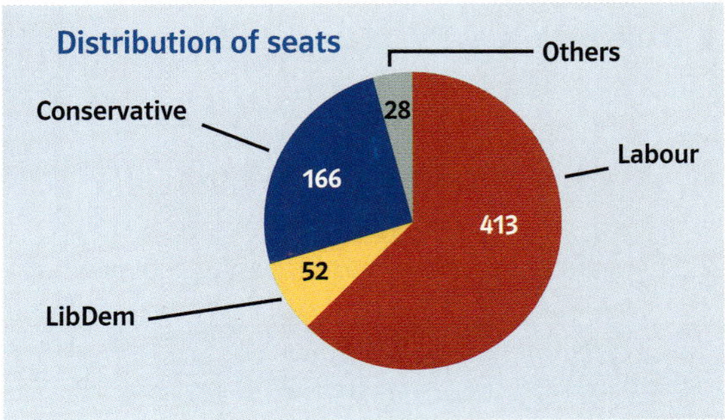

2 The British parliamentary system

■ Describe the diagram of the British parliamentary system.

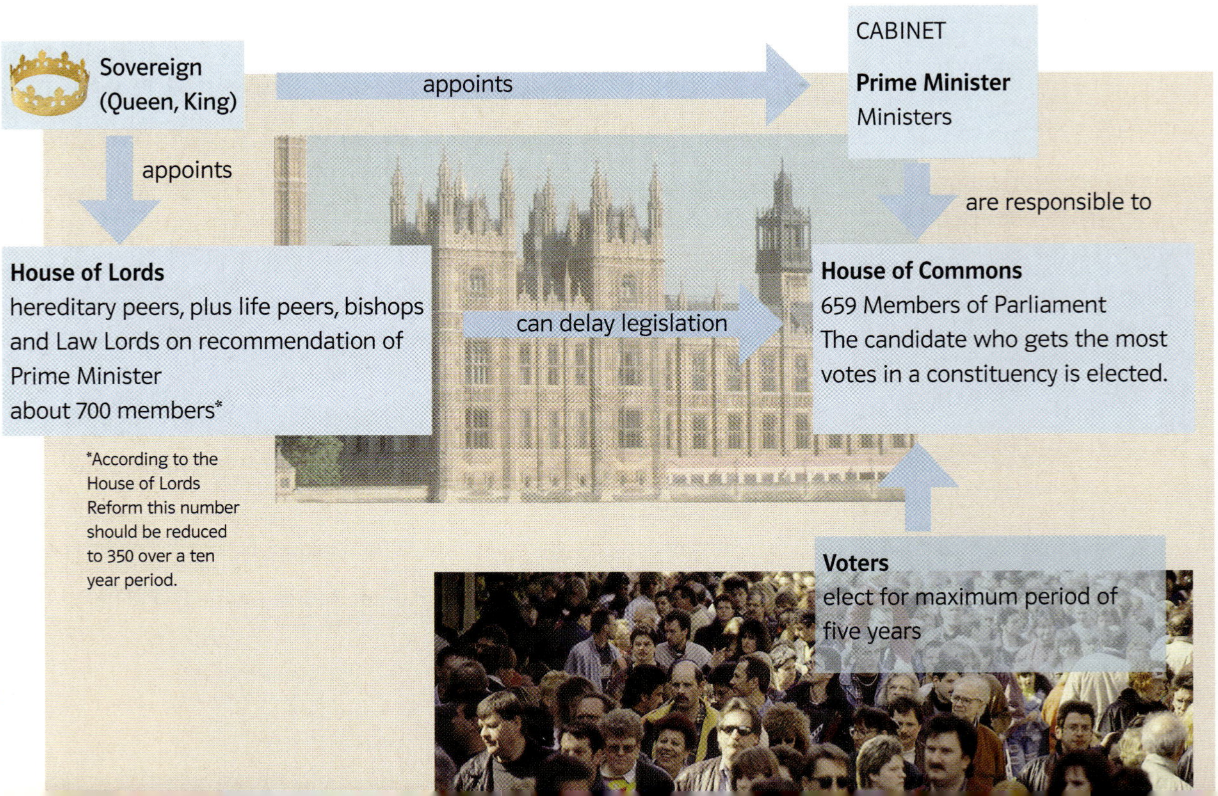

Unit 4

Telecommunications/Advertising

Answer the following questions.

1. What can you see in the pictures?
2. Do you own or use any of the devices shown?
3. Are they useful?
4. What advantages do they have?
5. What developments do you think we will have in the future? Think of office technology and communications.

1 A virtual future?

Since the invention of the computer technologists have promised us a world in which we will have more control over our lives – in short a technological paradise in the real world. Unfortunately this technology is sometimes still much too complicated and often not user-friendly enough. We can't even control the technology in our own homes; for example our PC systems still crash whenever we need them most.
But now a new trend in user-friendly technology is coming as engineers combine computers, consumer electronics and the media. Super appliances and services will change all our lives. Firstly, powerful computers are becoming smaller and smaller. Secondly, information transmission has gone digital. A new world of easy-to-use communication with higher speed data networks will open up for us all. People all over the world, not only the western world, will be able to send and receive multimedia documents that combine texts, sound and pictures. This computer revolution will be even bigger than the PC revolution of the 1980s. Why? The computer revolution of the 1980s reached only about 15 % of homes, mainly because people didn't really need expensive and complex machines to store addresses and telephone numbers, for example. The next revolution will introduce computer technology into most products that people need. The big difference will be that people will actually be able to use the devices because they will be more user-friendly. In future it will also become even easier to communicate with other users through wireless connection systems. Using digital devices people will be able to call up special computer programs. For example "VirtualTown" will then appear on the screen of your digital TV set. Here people will meet in virtual schools, make virtual trips to exotic countries, watch a football match and even talk to other users while they are watching. Of course, some virtual activities like shopping or banking via the Internet have already become reality and in future more people will make use of this new technology.
So we will all get the technology – and the future which we always wanted. It still won't solve the big problems, such as war and poverty. But at least we will have powerful devices that we will be able to use and not ones we do not really understand.

377 words

a Match the words in the box, according to the text.

networks • multimedia • wireless • systems • transmission • digital • documents • data • information • connection • PC • devices

b Answer the following questions.

1. Do we live in a "technological paradise"? Why (not)?
2. Which two developments have led to the new computer revolution?
3. What do we mean by "multimedia documents"?
4. Why will this new development be more important than the PC revolution of the 1980s?
5. How will "VirtualTown" work?
6. What big problems won't this technology solve?

c Find the words or expressions in the text from these definitions and then make four new sentences in which they appear.

1. electrical devices you use at home, e.g. a DVD player
2. apparatus for receiving and showing television transmissions
3. a complex system which connects and transmits information
4. a device used in the home or office which stores and processes information

d From the information in the text explain the technological revolution in your own words.

e In your opinion, what else could people use computers in their homes for?

f "We will all get the technology which we always wanted." Do you agree?

2 Who is in touch?

The bar graph shows the percentage of electronic communication devices in five European countries.

(2001) households with

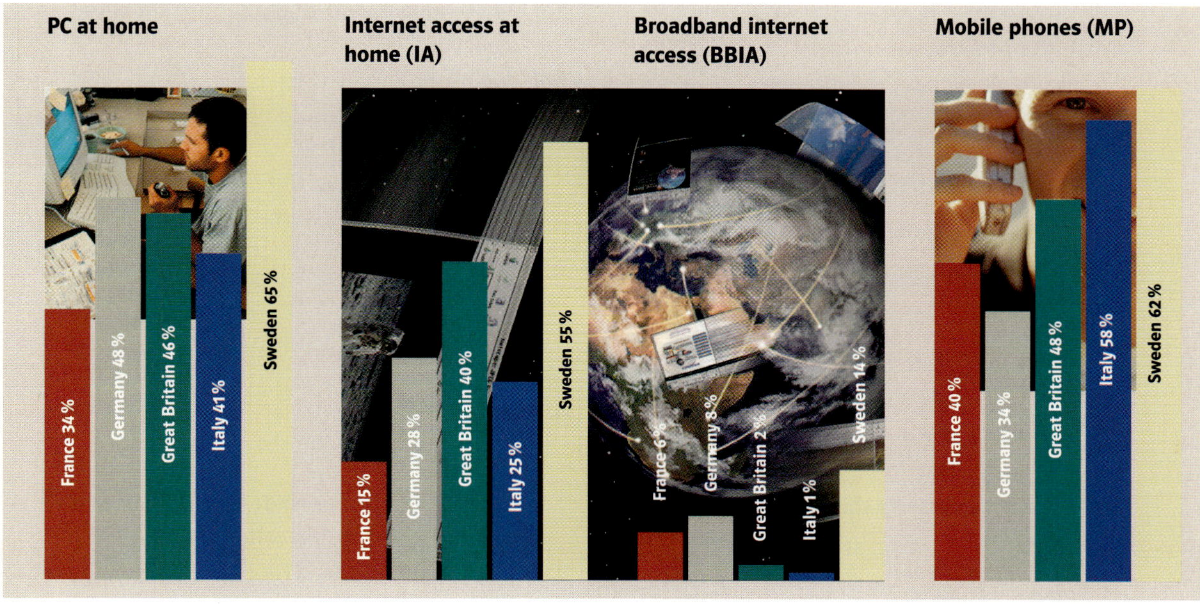

a Compare countries in the following way. Use the phrases in the box on the right.

Example:
IA – S / GB
The graph shows that the percentage of homes with internet access in Sweden is higher than it is in Britain.

1. PC – F / G
2. BBIA – I / S
3. MP – GB / G
4. IA – F / I
5. PC – S / F

The graph shows The figures show We can see from the statistics	that	the number of the percentage of	(device) owners households who have	is higher is lower	in (country A) than it is in (country B).

b What could be the reasons for these differences? Use the following phrases.

I'm pretty sure that …
That may be because …
A possible reason is that …
I'm sorry, but I've no idea why …

1 Life in the future – what will it be like?

a Ask your classmates what they think about the following ideas about the future.
The words in the box below can help you.

Example:
unemployment / increase
What do you think, will unemployment increase?
I think unemployment will increase.

I'm of the opinion; I think/don't think;
I expect; I suppose

1. people / read / books any more
2. computers / control lighting, heating and cooking
3. we / have / more wars
4. television, DVD, and computer games / replace / sport
5. there / be / more pollution
6. all houses / use / alternative energy

b Now answer the questions in short form and give reasons.

Example:
What do you think, will unemployment increase? Yes, it will because computers will do all the work.

2 In the office of the "Daily News"

Make sentences like this:

Example:
The phone is ringing. A secretary is walking towards it. She is **going to** answer the phone.

1. Jim Price has sat down at his computer. His notes for a story are in front of him.
2. Sally Jones is looking at her address book. She picks up the phone.
3. Two secretaries have left their desks and are putting money in the coffee machine.
4. Sam Smith, the head reporter, has invited a local politician to the office. He is preparing some questions.
5. It is eleven o'clock in the morning. Jim is taking some biscuits out of his bag.
6. The receptionist and her friend are in the restaurant. They have chosen their food and are standing at the till.

3 Office Electronics Ltd

Sam Smith wants to write an article about developments in the modern media for the "Daily News". He has invited James Bradshaw, manager of "Office Electronics Ltd", for an interview. Here is part of their conversation.

Put the verbs into the correct form of the future.

Sam Smith: Come in, Mr Bradshaw, sit down. **I'll take** your coat and I expect **Janet will** bring us some coffee in a minute.
James Bradshaw: Thank you, Mr Smith.
Sam Smith: Now, as I said on the phone **I'm going** to do an article on media developments and I'd like to ask you a few questions. Don't worry about that noise, by the way, it's one of the old fax machines. **It's going** to break down any minute.

James Bradshaw: Oh dear, I hope it's not one of ours.
Sam Smith: Oh no, of course not. Now, Mr Bradshaw, how is business? Do you think this year **1. (be)** as successful for you as last year?
James Bradshaw: Well, I certainly hope so. I have our most recent sales report here and the figures show that the firm **2. (do)** very well again this year.
Sam Smith: Of course, the market is growing all the time, isn't it? In the next few years I expect we **3. (see)** an even bigger increase in the use of communications technology. But do you think this **4. (be)** a good thing?
James Bradshaw: Well, for me and my company certainly!
Sam Smith: But what about the customers? It's an expensive business, isn't it? I've certainly decided that I **5. (not spend)** any more money on the latest developments.
James Bradshaw: Well, just a minute. I **6. (show)** you our brochure ...

4 The trainee journalists

Two new trainee journalists have started work at the "Daily News" newspaper in London. They are on a one month test period. Ken Brown, the head of the foreign news department, has to write a report on them. He has used the following system:
++ excellent / + good / – bad / –– terrible

	General work	Computer skills	Shorthand	French	Communication skills
Duncan	++	+	–	++	––
Tina	+	–	+	–	++

a First make sentences about Duncan. Use the verbs in the box.

speak – use – communicate – write – work

Example:
Duncan's general work is excellent. He works excellently.

b Form adverbs from the following adjectives.

careful, slow, fast, accurate, polite, quick, correct, hard, wonderful

c Now choose from these adverbs to compare the trainee journalists in the following way:

Example:
general work
Duncan works more carefully than Tina does.
Tina doesn't work as carefully as Duncan does.

5 A report **Advanced**

Look at the adjectives and adverbs. Then complete Ken's report on two other trainees.

On the whole I can say Phillip's work in the office has been **fairly good**. **Unfortunately**, he speaks French **terribly**. On the other hand, his computer skills appear to be **very good**. In comparison, Charlotte has worked **extremely well** over the last four weeks. Although she speaks French **(real) (bad)**, her computer skills are **(extreme) (good)**. Nevertheless, she seems to be the **(good)** of all our trainees. **(General)**, she keeps **(calm)** in a crisis and doesn't get **(nervous)**. All in all, I think she will **(probable)** make an **(excellent)** journalist. But she must learn to speak more **(polite)** on the phone. I **(certain)** cannot say the same about ...

Development of advertising

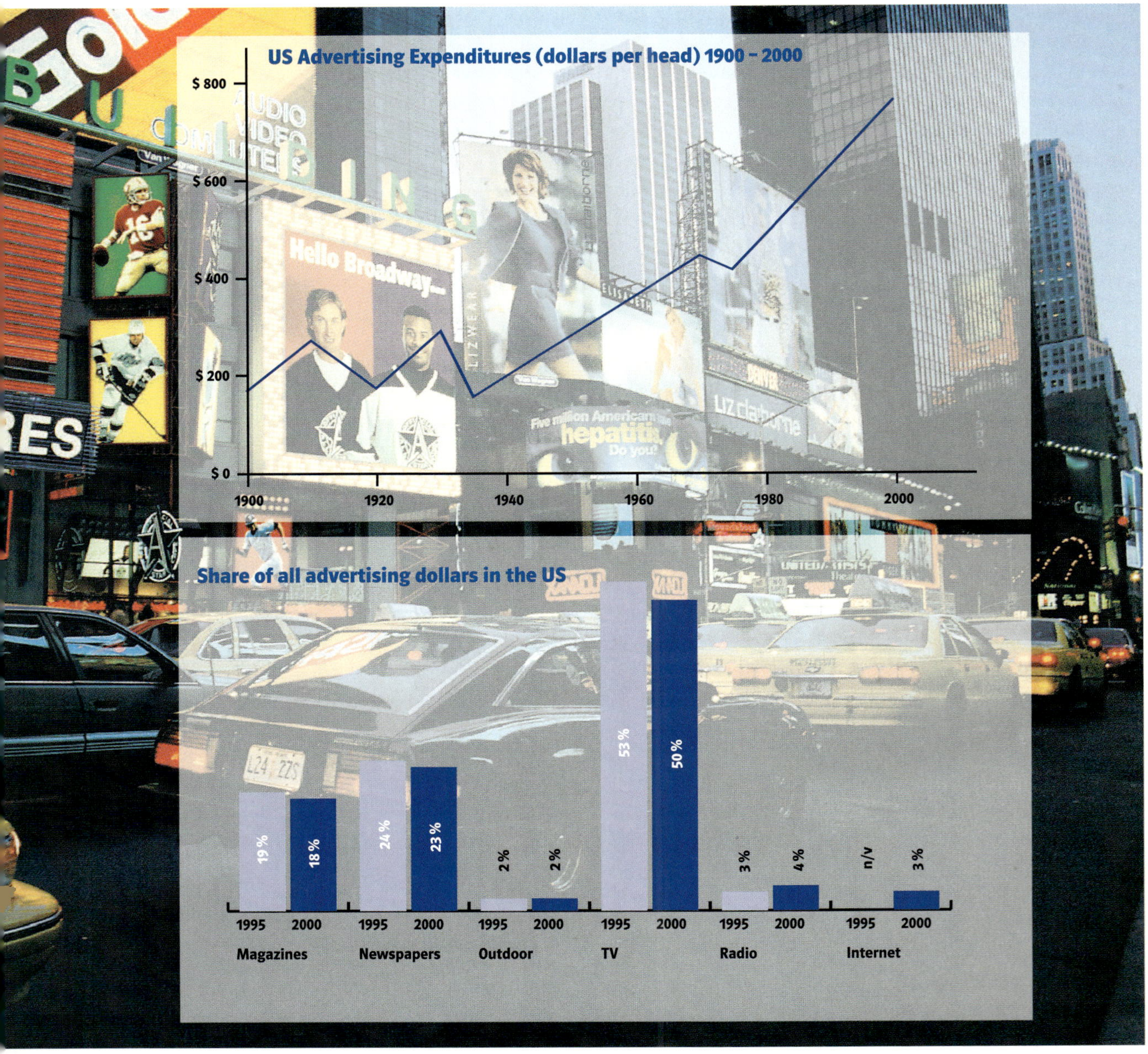

a Answer the following questions.

1. How did the amount of money which was spent on advertising per head change in the 20th century?
2. Whose share of the advertising market fell between 1995 and 2000?
3. Which of the media had a bigger share of the advertising market in 2000 than in 1995?
4. What could be the reason for this development?

b Read the text on the next page and answer the following questions.

1. Which of the media in the chart above are mentioned?
2. What could the headline of the following text be?

If you see a big yellow M on TV, will you recognize it and – more importantly – will you know what product it stands for? Most people will recognize it, according to an international survey. They believed that the M was one of the best known logos in TV advertising.

For the companies that had spent millions on the production of commercials and on airtime the results of the survey were quite important. The companies whose products came first were pleased. But the problem is that compared to every commercial that becomes popular 100 others fail.

Indeed, according to the survey, one of the main disadvantages of TV advertising is that only a third of commercials receive our full attention. One in five plays to empty rooms, while people are making tea or going to the toilet. Two in five act as background while people are eating, reading or even sleeping. With the help of "hidden" video cameras in a number of homes, the survey showed what surprising things people actually do during the break – one man even practised his tennis serve.

According to the survey even people who stay in the room during commercials do not always watch them. A new generation of viewers has arrived – people who watch with remote control in hand and switch to another channel when the commercials come on. Reports from America, where this habit is much older, say that in this way companies waste a quarter of the money which they spend on TV advertising.

Of course, it was not the TV companies which paid for this international survey, it was the newspapers. They used it for an £8 million advertising campaign. The message was simple: Advertisers overestimate the value of TV commercials. The reason why newspapers started their campaign is also easy to see; their share in the advertising market has been decreasing worldwide, and is down to about 15% in Britain.

Newspapers and TV stations have different viewpoints on advertising. On the one hand, Bert Hardy, who is the manager of the newspapers' campaign, argues: "The most expensive time in TV advertising costs around £60,000 a minute, whereas a full page in a national newspaper only costs £20,000 and lasts all day. The companies could use some of the money that they waste on TV advertising to increase their profits and cut prices." On the other hand ITV manager Malcolm Wall points out the fact that newspapers themselves spend millions on TV advertising.

Today many advertising managers are sure that a mix of the advertising media will be the best way to reach consumers. Some companies have already changed their advertising campaigns. Many more are planning the same and are going to advertise not only on TV in the future but also on the radio, in newspapers and on the internet. **465 words**

(The Telegraph plc, London)

c **Find the expressions with the same meaning in the text.**

1. 20 per cent
2. 25 per cent
3. 33 per cent
4. 40 per cent

d **Try to make some more expressions in the same way.**

1. 66 per cent
2. 75 per cent
3. 10 per cent
4. 80 per cent

e **Read the text again and answer the following questions in complete sentences.**

1. Are all companies happy with the results of their TV advertising? Why (not)?
2. What percentage of the commercials on TV do viewers really watch?
3. What do people do during the commercial break?
4. Who paid for the international survey on TV commercials? And why?
5. What financial advantage do firms have when they advertise in newspapers?
6. What media will many companies use to advertise in the future?

f **Give your opinion.**

If advertising on TV is so expensive, why are firms willing to spend so much money on it?

g **Translate lines 40 to 51.**

1 Radio advertising

a Listen and find out:
1. the product
2. the name of the product
3. the price
4. the name of the shop that sells it

b Listen again and find adjectives from the ones below which describe the products. Say which of the products they describe.

careful • cheap • superfast • superb • modern • safe • excellent • popular • polite • wonderful • light • accurate • good • quick • easy to use • comfortable

2 Analyzing advertisements

a Try to analyze the advertisement on the left.
1. Say what this advert is for.
2. Identify
   - the slogan/key words.
   - the logo/brand name/name of the company.
3. Say what other information is given in this advert.
4. Describe what the picture shows.
5. Say what ideas and lifestyle you associate with this picture.
6. What target group could the advert be aimed at?

b With reference to what you have found out above do you think this advert could be successful?

3 Writing an advertisement

a Look at the photo on the left and think of a product/service you could sell with it. Choose a suitable target group.

b Now write a short advertisement for your product/service. Refer to the points in D2.

c Present your advertisement to the class. Say why you have designed it in this way.

Unit 5

Transport

a Describe the situation in the photo.

b How do the people in the cars feel?

c Decide which of the following comments in your opinion best describe the situation in the photo and say why.

1. People in cities don't need their cars.
2. Private transport will be the death of us.
3. If you want to reduce your stress level, leave your car at home.
4. It's quicker on foot.

d What is your view of the traffic situation in our towns and cities?

1 Traffic

Every day we can hear reports of cities which are congested by traffic and this situation is getting worse day by day. The congestion of roads and motorways causes major problems for drivers and the environment. These problems have grown with the development of private transport. We have got used to a lifestyle we find hard to change. More and more people who own cars are not able to imagine a life without them. The question is how to find a solution people would accept. Some people believe public transport should be the answer. Others are looking into technological innovations such as electronic navigation systems. **109 words**

a Match the following words with their definitions.

1. congestion	A. the world around us
2. environment	B. a way to travel for people in large groups, e.g. bus, tram, train
3. navigation system	C. the state of being too full or overcrowded, especially on roads
4. public transport	D. a device which helps you find your way

Here are some statements we heard at a transport planning meeting.

b Complete these statements with the words used above (1.–4.).

1 *"During the rush-hour the ... on the M25 leads to long traffic jams every day."*

2 *"The fumes from cars pollute the ..."*

3 *"A ... can help drivers to avoid congested traffic without having to look at a map."*

4 *"I believe we should use ... and leave cars at home."*

2 Car navigation systems

Car navigation systems are no longer a high-tech dream. You can find them in many cars as a standard device. If you miss a traffic report, there is no need to worry. Your navigation system saves it for you in text and audio form and will give it to you at the touch of a button. It can even recalculate your chosen route using the latest traffic information. Life couldn't be easier!

Drivers can relax as they are guided to their chosen destination, be it a city centre or a small suburban street, without having to look at a single map. A quick look at the display and you can see how much longer your journey will take and how many miles you have to drive. All this is made possible by satellites that can pinpoint your car's location, if it is equipped with a Global Positioning System (GPS).

The Trafficmaster system is an additional aid in avoiding traffic congestion. Instead of relying on radio reports, Trafficmaster obtains its information from sensors installed on motorways and road networks which monitor traffic speed. Trafficmaster transmits this information to its users at regular intervals. The system is available as part of the navigation system from some car manufacturers, but is not normally a standard feature and you have to buy it as an extra.

So, is there anything navigation systems cannot do? Well, they cannot regulate traffic flow to avoid traffic jams and slow moving traffic in the first place. They can only help find a way out of the congestion and suggest a diversion. The other disadvantage with most navigation systems at present is the slow and rather complicated method of entering your destination, especially when you are driving. However, this is all going to change. A leading technology company and a car manufacturer have produced a navigation system that uses voice recognition. For example drivers who suddenly realize that they are running out of petrol can press the 'talk' button on their steering wheel and give commands like "find the nearest petrol station". The system is programmed to understand around 150 commands and a number of different accents. This makes programming your system a lot easier and safer, for example, when you are driving along the motorway or looking for the nearest fast food restaurant. However, all this only helps the individual driver and does not really solve the general problem of traffic congestion and pollution. **406 words**

a Look at the text.
1. Describe how you can find where the nearest petrol station is through a car navigation system.
2. Explain how this is possible.
3. Show how the Trafficmaster system works.
4. Point out two disadvantages of car navigation systems.
5. Explain why the newer navigation systems are safer.

b **Here are some radio travel news announcements. Find nouns in the text to complete them.**
1. Traffic … on the M1 near Nottingham is causing problems for drivers.
   Try to avoid the M1 near Nottingham if possible.
2. Due to motorway repairs on the M4 there is a 5-mile … in both directions near Bristol.
3. Road works on the A58 south of Rochdale. The police advise drivers to follow the … signs.
4. Due to an accident on the A30 near Exeter there is 3 miles of …
   So take care when driving on the A30 near Exeter.
5. Fog on the M62 ! Drivers should reduce their …

c **Rewrite lines 10–18 and replace some of the expressions with those in the cloud. Start like this:**

Example:
Drivers can take it easy as …

monitor · **take it easy** · route · glance · just one · find exactly · how far · fitted · position · shown the way

1 Getting around

a Look at these pictures and the information below. Then make sentences like this:

Example:

1. The train which is standing at the station is going to Docklands.
2. The students who are waiting for the ferry …

- go to Docklands
- go on a bike tour
- phone his wife
- be a London tourist bus

b Now describe the other pictures in the same way.

Example: The train which is going through the Eurotunnel …

- go to Paris
- travel to Italy
- read newspaper
- go to Paddington

2 Travelling home

Complete the following story. Fill in "who" or "which" where necessary.

At 5.30 Henry arrived at the station **(–)** he knew so well. He showed his ticket to the man **who** was standing at the gate. Then he got on the train **which** was waiting at platform 7. He politely greeted the woman **(1)** was sitting next to the door. She put down the book **(2)** she was reading. They both smiled and then he walked to the seat **(3)** he always chose. He sat down and filled the pipe **(4)** he had taken out of his jacket. This was the time of day **(5)** he liked best. After some minutes he started to read the paper **(6)** he had bought at the station. The couple **(7)** were sitting next to him were talking about the plans **(8)** they had for the weekend. He listened for a while and then he looked out of the window at the countryside **(9)** was passing by. What a nice day it was! Henry started to think about the work **(10)** he had done at the bank. He smiled. He remembered all the years **(11)** had passed and all the people **(12)** had worked with him. When the train arrived at the station he at once saw the woman **(13)** was waiting for him. He walked up to her and kissed her. It was the first time his wife had met him at the station. She had bought him the flowers **(14)** he liked so much and he knew it would never be the same again. This had been his last day at work after 48 years.

3 A guessing game

a Guess who or what it is.

1. It's a person who goes on a journey for pleasure.
2. It's an electronic appliance which helps you to find your destination.
3. It's a place where trains arrive and leave.
4. It's a piece of paper you buy when you want to travel by train or bus.

b Now describe other persons, places, vehicles etc. in the same way. Don't mention their names. Your classmates should find out who/what they are.

4 Changes in travelling

Compare travelling today, in the past, and in the future. Make sentences like this:

Example:

Lots of people must/have to wait in traffic jams today. 100 years ago most people had to walk.
In 100 years' time I think people will …

Today …	100 years ago …	In 100 years' time (I think) …
lots of people/wait/ in traffic jams (must/have to)	most people/walk	people/use/public transport in cities
passengers/travel/to America in a few hours (can/be able to)	passengers/get/to America in a week	passengers/travel/around the world in minutes
drivers/stop/at most European borders (needn't/not have to)	drivers/have/a driving licence	drivers/stop/at any borders
travellers/enter/some countries without showing their passports (can/be allowed to)	travellers/take/their cats and dogs into Britain	travellers/enter/all countries freely
man/fly/to Venus (can't/be not able to)	man/fly/to the moon	man/fly/to the stars
people/drive/ faster than 70 mph on British motorways (mustn't/be not allowed to)	people/drive/without a person carrying a red flag in front of their vehicles	people/drive/cars in city centres

5 A day in the life of Harry S. **Advanced**

Harry **(1)** works in London. But he lives in Bacon End **(2)**. Every day he has to travel one and a half hours to the computer firm **(3)**. At 7.30 when he leaves his house, he gets into the car and drives to the station **(4)**. Harry arrives there at about 8 o'clock and he gets on the train **(5)**. On the train he reads the newspaper **(6)**. Sometimes he talks to the passengers **(7)**. When he arrives in London he takes the underground line **(8)**. Sometimes he meets his colleague **(9)** and together they walk to work. On his journey home **(10)** he often uses his mobile phone to text his friends or listens to his CD player.

Complete the text about Harry S. with the information below. Use relative clauses.
Decide if the relative clause is necessary or not and find out if you can replace "who" or "which" by "that".

Example:

Harry, who is 24, works in London. But he lives in Bacon End, which is 35 miles away.
Every day he has to travel one and a half hours to the computer firm he writes programs for. At 7.30 …

Some more information on Harry S.:

a) The journey home takes about one and a half hours.
b) The line goes to Cannon Street Station.
c) Harry is 24.
d) He writes programs for the computer firm.
e) He usually buys a newspaper at the station.
f) Bacon End is 35 miles from London.
g) The passengers are sitting next to him.
h) His colleague also works at the computer firm.
i) The station is in Harlow.
j) The train leaves at 8.05.

1 From jam to tram

1 Peter Wood used to travel to his office in Manchester by car. Every morning the journey took longer and longer because the roads were too congested. One day he decided to try the tram and the journey was half an hour shorter. Thanks to the town's tram system, his journey only takes 16 minutes. "It really is a very good service," says Wood. Other commuters are happy too. One of them, Mike Pulman, adds "I stopped using my car years ago simply because the tram is so convenient."

2 Their views of the tram are quite different to the ones people in Britain had after the war. At that time people thought that the trams were noisy and slow and so they got rid of them. In the 1920s, Britain had over 14,000 trams in more than 100 towns, but by the 1960s the tram remained in only one town. At that time the attractive alternatives were the bus and the car. The bus because it provided a more flexible alternative to the fixed-route tram and the car because of the freedom it gave to the individual driver. Of course, in those car-crazy days, "nobody thought about the congestion and pollution which all those petrol engines would produce," says Don Roach, who is head of London's traffic department. Today however, environmentalists and city planners are welcoming back the electrically-driven, new-style tram or 'light-rail system' as they now call it. They see it as an efficient and ecological solution to inner city traffic chaos.

3 Today's trams are most profitable in cities which have populations of half a million or less and no more than 20,000 passengers an hour at peak times. Although one alternative – an underground system – may be a good solution to the traffic problems of large cities, Mary Browning, a British environmentalist says "it doesn't make sense for medium-sized cities and is much more expensive to build than a tram system." In fact, a team of traffic experts has estimated that building one kilometer of underground costs 11 times more than building the same length of tramway. "The main reason trams are coming back is not just because they are good for the environment," says Browning, "it's because they are cheaper and more efficient."

4 However, the fact that there are two good arguments for trams doesn't mean that they are easy to introduce. Public transport of any sort is unpopular with car drivers and the car industry so that governments do not always support it. As a result, it is difficult for town planners to put tramway projects into practice. This helped to slow down the opening of Manchester's system by nine years and may also slow down progress on similar projects. However, since statistics show that there will be twice as much road traffic by the year 2025, sooner or later governments will have to look at light rail systems as one possible solution to the transport problems which European cities are facing in the 21st century.

498 words

a Find the forms of transport mentioned in the text.

b Match four of these headings with the paragraphs in the text.

Trams – an old idea made new • Trams – a danger to private transport? •
Trams reduce journey time • Trams – too noisy and dirty •
Trams save money for medium-sized towns

c Now do these tasks.

1. Describe the difference which Manchester's tram system has made to commuters.
2. Say why trams were unpopular in Britain after the war.
3. Explain the advantages which today's modern 'light-rail systems' have.
4. Compare the costs of a tram system with an underground system.
5. Give reasons why it is often still difficult to introduce tram systems.

d Complete these sentences with information from the text. Use your own words.

1. As a result of road congestion in Manchester commuters …
2. Traffic planners are in favour of the tram because …
3. Although Britain had 14,000 trams in the 1920s …
4. Buses became more popular than trams because …
5. Although there are two good arguments for trams …
6. As a result of the increase in road traffic in the future …

e Translate lines 11 to 31.

2 Solutions to traffic problems

Two traffic experts are talking about possible solutions to traffic problems:

A: But **don't you think the best thing would be to** improve the public transport system?
B: Certainly, but that's not enough. People will never give up private transport completely. They would lose their freedom and industry would suffer. **I think the answer to the problems of** congestion for example, is technology. Why not make car computers cheaper, then everyone could have one?
A: Yes, but that doesn't solve the problem of pollution, does it? **One solution to that problem could be** a tram system. It's the quickest and most convenient means of transport and **I would definitely suggest it** for a medium-sized town.

■ Discuss the solutions you can see in the pictures below in the same way and any others you can think of. Use the expressions in bold type in the dialogue or similar ones.

1 Travel announcements

a Listen to the announcements. Where could you hear these announcements?

b Listen to the announcements again and answer the following questions.

1. Where is the train to Liverpool?
2. Where does it stop?
3. What must passengers do who want to travel to Leeds?
4. Where is the plane with flight number BA98 flying to?
5. Which gate should the passengers go to?
6. What was the reason for the delay?
7. Why is there trouble on the M4?
8. How long is the traffic jam?
9. According to the police, what should the motorists do?

2 Welcome to Wakefield

It's so simple to get here!

Since the great coaching days of the 18th century Wakefield has always been at the "crossroads of the north". This fact is now even more important in a world where fast and easy communications are a must.
In any part of the Wakefield district you are only minutes away from a motorway, the M62 to the two main ports of Hull and Liverpool, and the M1 to London which is only a three hour drive away.
Moreover, you can reach Wakefield by rail from all parts of the country. Regular train services operate from most cities in England, Scotland and Wales. Again London is only two hours away.
On top of all this, Wakefield has two major airports nearby: Manchester (only an hour's drive on the M62) and Leeds (only minutes away via the M1). These fast growing airports provide links with four major intercontinental airports – London, Amsterdam, Paris and Dublin.
An ideal base. Discover Wakefield yourself. It could work for you.

For further information write to:

Wakefield
Tourist Information Centre
Town Hall
Wood Street
Wakefield WF1 2HQ

Or visit our website:
www.city-visitor.com/wakefield

162 words

a With the help of the text and the map answer these questions about Wakefield.

1. Why do you think Wakefield is called the "crossroads of the north"?
2. How can you get to continental Europe from Wakefield?
3. Why are the Manchester and Leeds airports so important for Wakefield?
4. Why do you think Wakefield has produced this promotion brochure?

b Design a promotion brochure in English for your town.
Concentrate on road/rail/air/sea connections.

Environment

Waste management

In one year the average family of four throws away 112 pounds of metal, 90 pounds of plastics and an amount of paper equal to six trees. We produce more than 25 million tons of waste each year. Where does all this waste go?

Waste Management in Europe (per cent)

Country	Landfill	Energy recovery	Composting	Recycling
Denmark	11	58	2	29
France	49	39	6	6
Germany	54	18	5	23
Italy	80	7	10	3
The Netherlands	12	42	7	39
Spain	74	6	17	3
UK	85	8	1	7

a Describe what you can see in the picture.

b Answer the following questions.

1. What kind of waste do you think it is?
2. Where, in your opinion, does this waste come from?
3. What will happen to the waste?
4. What else could they do with it?

c Talk about the information in the chart on the left.

d What are the current developments in waste management in Europe?
Compare the latest figures (e.g. from the internet) with the figures in the chart.

1 Everyone can help to save the world

1 We are all environmentalists now. We all know about acid rain and the ozone hole. But what are we doing about these problems? As consumers we have the power to make decisions which will contribute to a better world. For example, when we walk into a supermarket we can choose to buy environmentally-friendly products and refuse products which damage our planet. In this way we also have the power to influence retailers to sell us these so-called 'green' products.

2 The supermarkets have understood the message and are now trying to improve their environmental image. They are turning to less wasteful and recyclable packaging. The range of 'green' products will continue to expand to meet the growing demand. This message has also reached the manufacturers. Companies are now developing environmentally-friendly products which mean no decrease in quality or increase in price. They have realized that protecting the environment may start as an advertising campaign but in the hands of intelligent people it can become a very profitable business.

3 We also know there are other areas where all of us can make a positive contribution to a cleaner, healthier and safer environment. We can all reduce energy consumption at home and so cut the emissions from power stations. There are a number of ways we can do this. For example, we now know how important it is to insulate our houses well. In the use of electrical appliances and lighting we should choose the best energy-efficient technology, such as modern light bulbs which can reduce electricity consumption by 80% over the older ones. Rubbish is another area where we can act. We all produce tonnes of rubbish and so waste disposal is becoming a big problem. As an environmentally-conscious consumer we should buy products which do not have so much packaging. We should avoid non-returnable goods at all costs. In Scandinavia and Germany, for example, plastic bottles and cans are now returnable goods.

4 In conclusion it is important to emphasize once again the rise of consumer power. We must realize that we are all part of the problem. Nevertheless we must also realize that we have the responsibility to be part of the solution.

363 words

a The text above has four paragraphs. Find headings for each paragraph in the following bubbles.

- The influence of consumers on supermarket products
- The use of public transport
- The introduction of 'green' products and price increases
- The problems of acid rain and the ozone hole
- Everybody's duty to contribute to a cleaner world
- The reaction of retailers and manufacturers to environmental problems
- Other ways to protect the environment

b Answer the questions on the text.

1. Which decisions can a consumer make in a supermarket?
2. How have supermarkets reacted to these decisions?
3. Why have manufacturers become more interested in environmentally-friendly products?
4. How can we reduce our consumption of energy?
5. How can we help to avoid waste?

c Give your opinion.

1. What is the message of the text?
2. In what other ways can you protect the environment?
3. What other examples of 'green' products can you think of?

d Look at the following nouns and find the verbs in the text. Say which line in the text.

1. choice
2. reduction
3. insulation
4. production
5. emphasis
6. sales

e Now find the nouns from the following verbs in the same way.

1. to consume
2. to demand
3. to decide
4. to contribute
5. to waste
6. to solve

f Find synonyms in the text for the following words.

1. shop owners
2. 'green' products
3. waste
4. to manufacture
5. to cut
6. to make better

2 Pollution

a Read the statements and match them with the photos.

A Emma Shaw, chemist: "I've just read a book about acid rain. Did you know it was a Scottish chemist, Robert Angus Smith, who invented the expression 'acid rain' in an 1852 article on the air and rain in Manchester? Moreover, according to him air pollution damages buildings and forests more quickly in countries where people use more fuel than elsewhere."

B Paul Jenkins, environmentalist: "It's true that this government is doing a lot to reduce the level of air pollution here in Britain. However, in my opinion, we consumers have the power to make the biggest improvement. We should buy cars which don't pollute the environment so badly or we should use public transport."

C Jill Kemp, journalist: "As I see it, we, the people of Britain, have been using our rivers as waste disposal channels for hundreds of years. I'm convinced that our government must act. Furthermore, we need better pollution control over companies, farmers and even local government."

D Richard Palmer, politician: "I think the ozone hole is the greatest danger to our planet. It isn't true that we can afford to wait. The ozone hole will cause serious dangers to our health. In addition all over the world people must stop making household products with CFCs* in them."

* CFC = chlorofluorocarbon

b Why did you make those decisions? Answer in the following way:

Example:
I chose the photo of … for text A because Emma Shaw says/tells us …

c Report what each person says and introduce each reported sentence with a different verb. Use the table below.

	Direct speech	Reporting verbs
Opinion	"In my opinion …" "As I see it …" "I think/believe …"	be of the opinion take the view think/believe
Emphasis	"I am (I'm) convinced …" "We should/ought to/must …"	maintain emphasize
Agreement/ Disagreement	"It is (It's) true …" "It is not (It isn't) true …"	agree not agree
Additional Information	"Moreover …" "Furthermore …" "In addition …"	point out go on to mention add

1 Protect your environment

For a project on the protection of the environment, the consumer magazine 'What?' interviewed some people in Birmingham. Here are some of the answers they got.

1. "We are fighting for more cycle paths here." **Joe Straw, member of Green party.**
2. "I'm joining the WWF tomorrow. People don't give enough money to help with their work." **Roger Williams, bank clerk.**
3. "I don't buy cans." **Mrs Black, housewife.**
4. "I ride my bike to school every day." **John Smith, teacher.**
5. "We are closing some streets to traffic in the city centre next year." **Ray Peters, town planner.**
6. "My husband doesn't wash his car here in the street any longer." **Mrs Rivers, housewife.**
7. "We always buy bottles we can return to the shop." **Fred and Jim, college students.**
8. "Well, I'm just taking my old newspapers to the container like I usually do on Saturdays." **Peter Bone, old man.**
9. "We don't buy goods which aren't environmentally-friendly." **Anna and Rob Nelson, young married couple.**
10. "We must all save a lot more energy in the home. We're buying an energy-efficient heating system tomorrow." **Mr Field, farmer.**

■ Report the answers like this:

Example:
1. Joe Straw, a member of the Green party, said that they were fighting for more cycle paths there.
2. Roger Williams, a bank clerk, said that he was joining the WWF the next day. People didn't give …

2 Retailers and the environment

The manager of a large supermarket group gave the following statement to 'What?' magazine about retailers and the environment:

"We opened our first supermarket just ten years ago and we have had a lot of success since then. We have always kept our prices low and our quality high. We will of course continue to do this in the future. However, and I know you are interested in this, we have always had a special interest in environmental matters. Right from the start we sold goods with as little packaging as possible and placed bottle banks and containers in all our car parks. As from next month we won't hand out free plastic bags any more and we will stop selling all drinks in plastic containers. Over the last few years we have introduced more and more 'green' products and we certainly haven't felt sorry about this. Two years ago, when we started to plan our second store here in this area, there wasn't a bus service. But after we wrote to the council they agreed to introduce one. All in all I think I can say that we are really doing our best for the environment."

Chris Hawkings, who reported this statement for the magazine, began his article like this:

Example: The manager of Dippy Supermarkets spoke to me yesterday … He told me that they had opened their first supermarket ten years before and that they had had a lot of success since then. He went on to say … that they had always … and that they would of course …

■ Complete his article.

3 Interview with "Whiz Bang Electronics"

Chris Hawkings' next job was to interview the head of "Whiz Bang Electronics" about their new washing machine, which they want to sell as an environmentally-friendly model. Here is his list of questions:

1. Do you plan to replace all your old models with environmentally-friendly ones?
2. How many models will you produce per year?
3. Why do you call it environmentally-friendly?
4. Do you think the new model will sell well?
5. When did you begin to develop this model?
6. How many litres of water does it use?
7. What is the energy consumption like?
8. What plans have you made to recycle these washing machines?

Later at lunch, the head of "Whiz Bang" told one of his colleagues about the interview and that the reporter had wanted to know a lot about their new model.

Example:

1. He wanted to find out if we planned to replace all our old models with environmentally-friendly ones.
2. He also wanted to know how many …

■ Now you go on.

4 The inner city traffic situation **Advanced**

Last night there was a meeting in the town hall to discuss the inner city traffic situation. Here is part of this discussion:

Town planner: … and so in my opinion, the only solution is to close the centre to traffic completely.
Shop owner: What will happen to our trade? No one will want to come into town at all then. And we already have enough competition from these big out-of-town superstores.
Town planner: But we have already introduced a number of highly successful park and ride schemes and I'm sure people will prefer shopping without all the noise and exhaust fumes from cars.
Shopper: That may be true, but we don't like carrying heavy bags of shopping for miles and miles before we can get them into the car.
Shop owner: You see, we must discuss the problem in more detail. Do you realize the problems we will have with deliveries?
Town planner: We do, Mr Black, and as long as lorries can deliver before 10 in the morning, there won't be a problem.
Shopper: I suppose shopping is more pleasant with no traffic. I will certainly feel better about bringing my children into town then. In some cities they even have special play areas for children. Can't we have something like that here, too?

■ Now report this discussion. Use as many different reporting verbs as possible.

Vehicles go round again

The three R's of waste management

Randy Newbury, manager at an international car company, has a dream. "In thirty years our company will be 100% environmentally friendly. On the one hand we will not pollute air or water any more during the production process, on the other hand we will only produce cars from recycled material. And I hope we will then be able to recycle these cars completely." This sounds unlikely? Not to Newbury, who is responsible for such environmental problems. What is more, lots of other international companies are actively trying to find solutions to make this dream come true. While they are already re-using waste heat and raw materials, they are reducing the amount of waste in the production process. Moreover, they are trying to increase the number of parts they can re-use. And of course, the products themselves must be recyclable. To be more precise, these companies are employing the new 3 R's to protect the environment: reduce, re-use, recycle.

The 3 R's were not always part of the companies' vocabulary. In fact, until the late 1980s, companies wishing to go 'green', installed filters in the factories and then tried to get rid of the waste they had collected there. "People were just shifting their pollution from one place to another," states Ron Humber for Friends of the Earth in London. "What we need is a completely new approach to waste management."

"I'm dreaming of a factory," says Randy Newbury, "in which scrap vehicles enter through the back door for 100 per cent recycling and leave as new cars through the front door."

"At the moment this is only a dream," he quickly adds, "but we must do something now. Every year millions of cars in Europe go to the scrap dealers."

"In the past, industry was only able to recycle the metal parts. But after they had taken them out they still had up to 250 kg of mixed waste – plastics, glass, rubber and engine oil. The scrap dealers had to pay £50 a tonne to dump this waste – and the costs were rising. Now special firms collect and dismantle old vehicles and return as many parts as possible to the car manufacturers.

We code the plastic parts of each car so it is easier to separate material for recycling. In this way we reduce the amount of mixed waste, but there is still some left. So our engineers and suppliers are working on this problem to find substitute materials for the rest. Ideally car parts should be made from one family of material. All this adds to the cost of a new car but, of course, we are able to save money by re-using or recycling the parts.

This means that one day we will produce a car which is 100% recyclable, which allows us to use the same materials over and over again. As we see it every new vehicle we design in this way helps us to solve tomorrow's problems today."

493 words

a Before answering the following questions on the text, look at the "steps" on the opposite page.

1. What is the manager for environmental problems dreaming of?
2. How are other firms attacking the waste problem?
3. In what way did companies try to solve the problems of pollution in the 1980s?
4. How can special firms improve the work of the scrap dealers?
5. How can car companies help to solve problems of pollution when they are designing new cars?

Follow these steps

Example (Question 1):

1. **Read the text carefully.**
2. **Find the lines in the text which contain the information necessary to answer the questions.** — lines 29 - 32
3. **Look up all the words (in a dictionary or below) you must understand to answer a question.** — scrap = metal from …

dealer (n) **1.** A **dealer** is a person who buys and sells things. **2.** A **dealer** in a game of cards is the person who gives out the cards to the other players.
dismantle (v) If you **dismantle** a machine, you carefully take it to pieces.
dump (n) A **dump** is a place where rubbish is left.
dump (v) **1.** If you **dump** something, you throw it away carelessly. **2.** If a company **dumps** goods, it sells large quantities at very low prices, usually in another country.
recycle (v) If you **recycle** things that have already been used, such as bottles or paper, you process them so that people can use them again.
reduce (v) To **reduce** something means to make it smaller in number, size, price etc.
re-use (v) When you **re-use** something, you do not throw it away. You use it again.
scrap (n) **1.** A **scrap** of paper is a very small piece of it. Usually you do not want it any more. **2. Scrap** or **scrap** metal is metal from old or damaged machines, cars, etc. which is melted so that it can be used again.
shift (n) A **shift** is a change from one position to another.
shift (v) If you **shift** something, you move it or change its position.
substitute (n) A **substitute** is a person or thing that takes the place of another.
substitute (v) If you **substitute** a thing, you put the new thing in the place where the old thing was.

4. **Use part of the question as an introduction to your answer, if possible.** — The manager for environmental problems …
5. **Use synonyms and paraphrases.** — The manager for environmental problems has a vision of …
6. **Write your answer in complete sentences. Be careful with the tenses and use the phrases below.** — It says in the text that the manager for environmental problems has a vision of a factory in which old cars become fully recycled new cars.

In the first (second, …) paragraph the author mentions that … / In line one (two, …) we read that … / It says in the text that …

b Translate lines 44 - 52.

Info-Box: How to answer questions on a text

Remember:

1. Read the text carefully.
2. Find the information necessary to answer the questions.
3. Look up all unknown words which are important for your answers.
4. Now answer the questions.
   - If possible, use the question as an introduction to your answer.
   - Use the main facts that you have found in the text.
   - Write your answers in complete sentences.
   - Use your own words.

1 Ecology or economics?

a Listen to the news report on Radio 4 and find the correct answers.

1. The news report is about
   a. a North Sea oil project
   b. a zinc ore mining project
   c. a coal mining project
2. The reporter mentions the effect of the project
   a. on the tourist industry
   b. on the air
   c. on the countryside
3. The people against the project are
   a. technologists
   b. hotel owners
   c. environmentalists

b Now listen to the recording again and find out more about the following points:

1. where the company has found the zinc ore deposits
2. why this area is especially interesting for the company
3. how many jobs the project will create
4. what this development will probably mean for nature in this area
5. the fears of the hotel owners

c Use your answers to write a short summary of the radio report.

2 Zinc UK project: A public enquiry

Zinc UK, a mining company, have found large deposits of zinc ore in the attractive coastal area of Cornwall. The ore is rich in zinc and is near the surface. This means mining will be relatively easy and therefore profitable. The project will create new jobs for the area.

The problems are that this is a particularly beautiful area and it is the home to many wild plants and animals. It is also a tourist attraction for the visitors to the small seaside resorts along the coast. The seaside resorts are small because of the poor road communications and there is no railway. The mining company has promised to build a new major road which will go directly to the motorway 25 miles away. Unemployment in the area is high. Most of the jobs are on local farms and in the seaside resorts. However, in the last few years the seaside resorts have had fewer visitors. Local taxes for house owners are high and the local government wants to keep tax increases as low as possible in the future. Zinc UK would contribute to the income of the area. The small town of St. Agnes has a new housing estate, but it only has a few shops, a coffee bar, a pub and a primary school. There is no public library, sports centre or cinema in the area.

228 words

a What advantages/disadvantages do you think the project could bring to the area?

b The local MP for North Cornwall has decided to hold a public enquiry to hear the views of the people concerned. The enquiry will help to decide whether Zinc UK should be allowed to start their mining project or not. The following people are at the meeting:

1. managers of the mining company
2. teenagers from the area
3. members of the local environmental group
4. hotel owners from the area
5. unemployed house owners from the area

Imagine you are one of these people.

1. List the advantages and disadvantages that the mining project might have for you personally.
2. Play your role at the meeting. Use expressions like "In my opinion" etc. to introduce your arguments.

c After you have discussed the issue write a report of the meeting. Use reported speech.

Energy

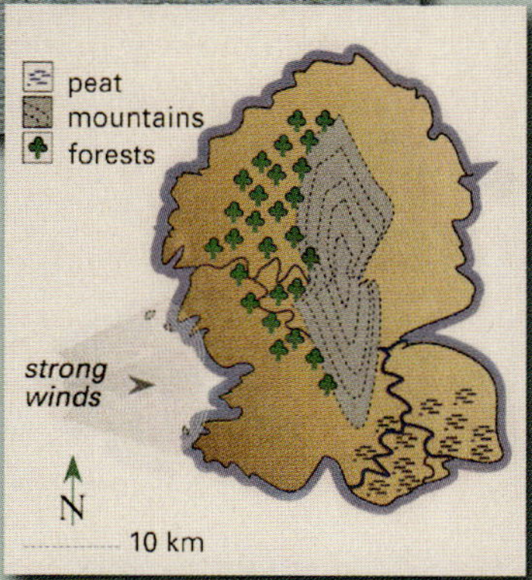

This is an island off the west coast of America. The island has strong winds from the west, high mountains with fast running rivers, thick forests along the west coast and peat bogs in the south east. There are no resources of coal, oil or natural gas on or near the island. At present only a few people live there.

You are in a project team which wants to develop the island. Think about and discuss these ideas.

1. What material would you use to build homes on the island?
2. Which two materials could help you to produce heat to cook food?
3. What could you use to produce electricity to run machinery etc.?
4. Where would you build houses on the island? Give reasons.
5. What environmental problems will you cause when you use the natural resources on the island?
6. What conclusion do you draw from your study?

1 More wealth – less energy

Modern technology and relatively low prices have led to a situation in which most people feel relaxed about energy. Technology has made us believe that there will always be an answer to future problems and the prices still create the impression that we have plenty of energy. But we have little reason to feel secure.

More than half of the world's population (mostly in the developing countries) lives without a commercial supply of energy. The total energy consumption per head in countries such as India is only about 3 % of that in the United States. By 2100 the number of people in the world will rise to about 10 billion (compared with more than 6 billion today) and 85 % of mankind will live in poor countries. Either billions of people will continue to live in poverty, or the demand for energy will grow enormously. Governments and companies will invest trillions of dollars in the next twenty or thirty years – more than ever – in the developing countries. And if they do so, economic growth will mean much higher consumption of energy in the Third World.

Although the availability of fossil fuels is limited, experts predict that during this century the world will still depend on them, especially coal which is plentiful in India and China – countries with the highest populations. The problem is that developing countries care far more about the standard of living of their citizens today than they do about pollution and global warming. That is why they use the little money they have to increase the production of goods. If they had enough money, they would be able to produce and use energy in an environmentally-friendly way.

On the other hand many wealthy industrialized countries are introducing methods to save energy and they are further developing technologies which provide new sources of renewable energy such as solar power. Of course, it is not an easy job to transfer these expensive technologies to the poorer parts of the world. But if we help the developing countries, they will be able to increase their energy efficiency. In this way these countries will manage to raise the standard of living significantly. At the same time they will consume only little more energy than today and the production of this energy will not harm the environment so much.

The time to take action is now. The West is beginning to learn from the mistakes it has made. Surely it is important to help the developing countries so that they can avoid the same mistakes. If we do not act now, what will life be like in 2100?

435 words

a Refer to the text and correct the following statements.

1. We must start to help the developing countries in the year 2100.
2. The number of people in the world is falling.
3. Today the United States consumes 5 times more energy per head than developing countries in South Asia.
4. In many developing countries people have developed methods to save energy.
5. The developing countries do not buy new energy technologies because they do not want to save energy.
6. The price of energy is very high today.

b Find the nouns in the text.

1. grow 2. consume 3. pollute 4. supply
5. demand 6. produce 7. poor 8. efficient

c Find antonyms in the text.

1. less 2. poor 3. high 4. a lot 5. past
6. industrialized (countries) 7. difficult
8. slighty 9. cheap 10. (the number will) fall

d Answer the following questions.

1. Which are the sources of energy mentioned in the text?
2. Which of them is a fossil fuel and which is a renewable source of energy?
3. Which of the following are fossil fuels, which are renewable sources: gas, water, wind, oil, peat?
4. Which advantages and disadvantages do fossil fuels and renewable sources of energy have? Think about these points: geographical position, availability, costs, pollution.

2 Talking about graphs

Graph 1: Development of world temperatures

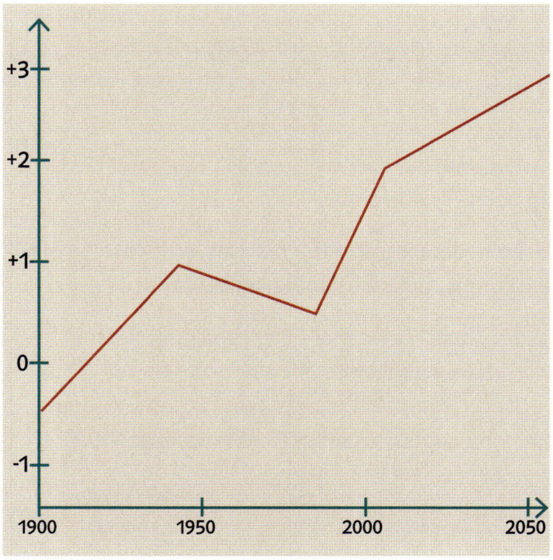

Graph 2: Development of energy consumption (1970 - 2020), in Quadrillion Btu (British thermal units)

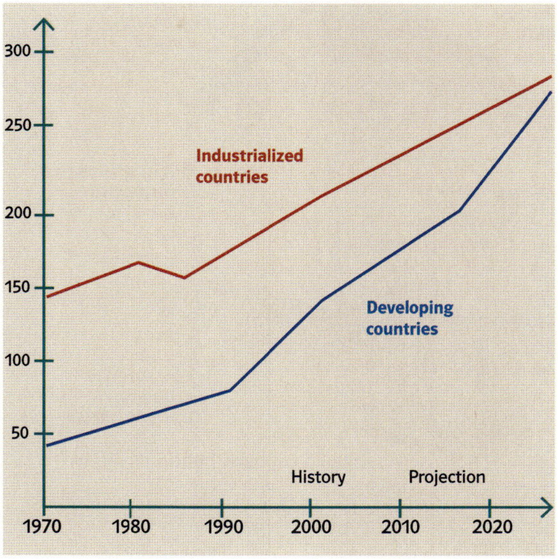

a **Study the graph above and its description.**

The graph shows the development of world temperatures.
Between 1900 and 1940 world temperatures rose significantly.
From 1940 to 1980 they fell slightly.
Since 1980 temperatures have increased constantly.
If this trend continues, temperatures will be about 3.5 degrees higher in 2050 than in 1900.

b **Now describe graph 2.**
Use the words in the boxes below.

Trends	
increase increase rise go up grow	**decrease** decrease fall go down drop
remain stable steady constant at the same level	

Changes	
big changes significant(ly) substantial(ly)	**small changes** slight(ly) a little
fast changes sudden(ly) sharp(ly)	**slow changes** slow(ly) gradual(ly)

1 How to save energy in your house

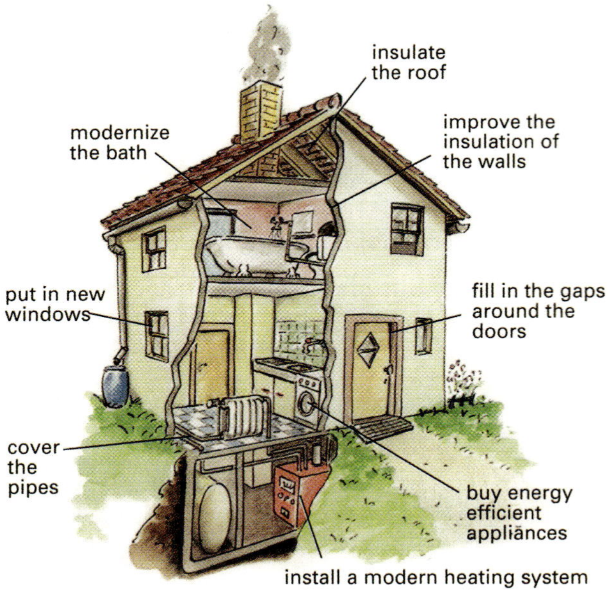

Darren and Kate Palmer have just bought a house. The house is 25 years old and needs some repairs. The Energy Council has sent them a brochure on ways they can save energy in their house. Darren is showing Kate the brochure and says: "Look, it says here …"

Example:
"If we insulate our roof well, we will save a lot of energy."

■ Use the drawing and the phrases in the box to make at least eight sentences in the same way.

lose less heat • save money • not use so much hot water • avoid a loss of heat • reduce our energy consumption • not pollute the environment so much • cut our water bill • not waste so much energy • save a lot of energy

2 The advantages of alternative energy

Cathy Johnson, a politician and expert on energy questions, is giving a talk at a local school. She is convinced that we depend too much on the use of fossil fuels (oil, gas and coal) as sources of energy. She believes that we should save energy and also look into the possibilities of the use of alternative energies more carefully. Here are some of her notes for her talk.

■ Use the expressions in the box to express her opinions.

1. First of all – must say	5. Moreover – must add
2. Secondly – be of the opinion	6. In addition – believe
3. Thirdly – think	7. Furthermore – must say
4. At this point – must emphasize	8. Finally – be convinced

Example:
1. First of all I must say, if people used alternative forms of energy, they would contribute to …

1. people use alternative forms of energy – they contribute to a cleaner world
2. government spend more money on these alternatives – we all see the advantages in a few years
3. electricity companies replace old power stations in this area – they can produce electricity and not pollute the environment so much
4. firms produce more energy efficient appliances for the home – we not need so many power stations
5. house owners insulate their houses better – they can reduce their electricity bills
6. we use the sources of renewable energy in this part of the country – we save a lot of fossil fuels
7. people not travel by car so much – we not have such a big energy problem
8. we not waste energy as we do now – we have a cleaner and better world

3 Bob Flynn is thinking about his life

Bob Flynn is the manager of a small company which develops solar systems.

■ Express his thoughts for him like this:

Example:
If I had worked hard, I would have got good marks in the first year. If I hadn't spent more time on my studies, I wouldn't have done well at college in the end.

1. I got a job as an electrician - after I wrote a lot of letters of application.
2. I went to university - after I attended evening classes three times a week.
3. I studied electrical engineering - after I looked at the good job possibilities.
4. I passed my examinations at university - because I worked hard.
5. I didn't go straight into my own business - because I wasn't sure what to do after university.
6. I started up my own business - after a friend gave me some tips.
7. I bought a workshop - after the bank lent me $50,000.
8. I became interested in alternative forms of energy - after I joined an environmentalist group.
9. I started to develop solar systems - because I learnt all about solar energy at university.
10. I didn't invest more money in the business - because I didn't have enough money at that time.

4 Save energy now **Advanced**

Two students have started a "Save Energy Now" group. They are talking about the use of energy and how people have reacted to new developments.

■ Put the verbs in brackets into the correct form.

Mary Jo: But we still aren't doing enough and if we **1. (continue)** to take so little care of our environment, soon we **2. (not have)** an environment at all.
John: I know. But if people **3. (know)** years ago about the damage that power stations can do to the environment, I'm sure they **4. (do)** something about it a long time ago.
Mary Jo: Yes, but now we all know about it and we can all help. If a lot of people **5. (not be)** as lazy as they are, they **6. (not waste)** so much energy in their homes. It's the little things that help.
John: Yes, you're right. I'm sure that if the school **7. (not help)** us with our project, then we **8. (never be able to start)** it, for example.
Mary Jo: That's true, and nowadays there are more people who are really interested in alternative forms of energy. If that **9. (not be)** so, we **10. (not find)** so many people at our meetings.

Renewable energy

In the mid-1970s fewer than 1,000 U.S. homes generated their own power. That number has grown to more than 100,000 and every year thousands more will join those who are already off-the-grid. They are turning to renewable energies, such as wind, water and sunshine in order to generate power for their homes.

But why are more and more people doing this? Very often it is because these people live far away from the grid and cannot afford the cost of a power line to their home.

As a result of this strong interest, it has become big business to supply homes with renewable energy. The yearly sales for the solar-energy industry alone have already reached $1 billion, and wind and hydro systems for domestic users add millions more. Recently many off-the-grid homes in the USA opened their doors to show how they keep the beer cold, the shower warm and the reading lamp lit with no help from the local power station.

Wind power is generally used for industrial and commercial applications. However, the most popular off-the-grid technology for private homes is the use of solar energy. Although photovoltaic (PV) systems are nearly twice as expensive as wind power, the sun provides energy for more than two thirds of America's off-the-grid homes. This is because solar energy is easy to install and doesn't need a lot of maintenance. In the past, makers of PV systems sold 60 % of their equipment to the government. However, their main customers now are private homes and companies who have realized that PV systems are often cheaper in the end. For example, private companies can install a PV cell on phones or on highway signs which are far away from the grid and so it's not necessary to dig up roads and lay cables. Even the building industry has shown an interest, offering solar-equipped homes at no extra cost.

One of the most important developments recently is that some electricity companies are even using solar power to avoid the need to build more traditional power plants. A Californian gas and electricity company has supplied 150 homes in California with electricity from a number of PV panels. In this way it saved the $ 1 million it would have spent on a traditional system.

"All this is going in the right direction," says Tom Kelly, a solar research scientist, "but we must still make a number of engineering improvements before we are ready for the mass market." In fact, research is going on all the time in order to find cheaper and better materials and to increase efficiency in solar energy installations. Off-the-grid energy may even get some help from politicians, who may introduce an extra tax on fossil fuels in order to limit global warming. If they do this, fossil fuel prices will increase and solar energy will therefore become more attractive. However, the energy of the future might come from a completely different source. Many experts believe that hydrogen-based fuel cell systems will provide the energy for the generations to come. Fuel cell powered cars are already on the roads and many companies are testing fuel cell systems. They are relatively cheap, renewable and environmentally-friendly. So, let's see what the future holds for us!

536 words

a Write a summary of this text.

Follow these steps

	Step	Example for first section:
1.	**Read the text carefully and write the main idea in an introductory sentence.**	In his text "Renewable energy" the author discusses the increase in the use of alternative energy sources, such as wind, sun and water.
2.	**Divide the text into six main sections and find headlines for each.**	Section I (lines 1–14) Off-the-grid energy increasing in U.S.
3.	**For each section write key points (not complete sentences). These must contain the main ideas in each section. Be careful not to include any unnecessary information or details.**	since 1970s – more U.S. country homes – off-the-grid power. alternative sources of energy – cheaper for them
4.	**Decide what function each section has, e.g. introduction to subject matter, presentation of arguments for and against, examples, further information, results, contrast and conclusion.**	Section I – introduction to subject matter.
5.	**Choose expressions from the list below which help you to present the ideas in the six sections.**	Firstly, the author points out that …
	Firstly / On the one hand / On the other hand / Furthermore / Consequently / As a result / For example / For instance / The reason for / Although / In conclusion / All in all	the author points out / claims / states / maintains / mentions / argues / adds / goes on to say / concludes that …
6.	**Now write a summary of the text using your key words and notes. It should be about half the length of the original text and you should use your own words.**	In his text "Renewable energy" the author discusses the increase in the use of alternative energy sources, such as wind and sun. Firstly, the author points out that it is mainly since the 1970s that U.S. country homes have turned to off-the-grid power. Alternative sources, says the author, are cheaper for them.

b Translate lines 58 to 78.

Info-Box: How to write a summary

Remember:

1. **Write the main idea of the text in an introductory sentence.**
2. **Divide the text into sections and find headlines for each one.**
3. **For each section write key points, giving the main ideas.**
4. **Decide what function each section has.**
5. **Choose expressions to present the ideas in each section.**
6. **Complete the summary in about half the length of the original text in your own words.**

1 From poultry litter to electricity

John James, an engineer from Britain, is on a business trip to Minnesota in the USA. He is talking to Chet Brown, production manager of an alternative power station which runs on agricultural biomass, including poultry litter.
The pictures below show some things which are part of the production process.

a Listen to their conversation and put the pictures in the right order.

turbine

furnace

fuel on conveyor

litter lorries tipping

b Now describe the production process.

2 Writing a report

Here are the notes which John James made during his visit to the Minnesota alternative power station.

1. generator produces 12.5 megawatts
2. fuel cheap and environmentally-friendly
3. government money still necessary to keep station competitive
4. lorries deliver 25 tons manure to storage hall daily
5. low manpower costs: 2 employees only
6. furnace burns manure at 800 °C
7. ash from furnace is quality fertilizer
8. turbine drives generator
9. greenhouse gas production 25% lower than from traditional power station
10. steam drives turbine
11. computer-controlled conveyor belts take poultry litter from storage hall to furnace
12. burning process produces steam

a From the notes make two lists to show:

1. the energy production process.
2. the arguments for and against the process

b Use John James' notes to write a report.

1. Explain the process in the correct order.
2. Write down points for and against the process in full sentences.

Automation

Answer the following questions.

1. What differences can you see between the two production methods?
2. When do you think the photos were taken?
3. Where would you prefer to work and why?
4. What changes in production methods will there be in the future? What do you think?

1 New production technologies

Modern life in the so-called civilized world needs all kinds of goods in increasing quantity and quality: computers, digital cameras, TV sets, toys, cars etc. There are many ways of producing goods, but for products that are sold in large quantities, the best way to make them is by mass production.

Manufacturers are constantly trying to improve their productivity. That is why automation has been introduced to work faster, more accurately and more economically. As a result factories and shop floors look very different from the way they did in the past.

a The following three paragraphs describe three important developments in new technologies which are used in modern factories.
Read the three paragraphs and find out from the box on the right which terms go with each paragraph.

robots, shop floors, Computer Aided Design (CAD), workers, research department, Computer Aided Manufacturing (CAM)

1 One of the latest developments in this area is that the "workers" are equipped with electronic sensors such as video cameras, microphones and touch sensors. Such a "worker" needs this equipment to do his work and also to send back information to the controlling computer. The computer is programmed to use this feedback to change its instructions to the "worker" if necessary.

2 With special programs designers can simulate on the computer what will happen if a particular design is chosen. In the past often unsuitable plans were drawn, just to find out later that they did not work. Now the computer helps to predict what will happen under different conditions – for example to a bridge in different winds. Moreover, these programs can also work out the most economical way of making something out of the materials available.

3 Machines that have to do the same job again and again can be operated automatically by this computer system. The computer programs in this system control the automated machines or tools. This software can be changed and so the machines can be programmed in different ways to make different products. The system is also used to monitor the production process and correct any faults.

b **Answer the following questions.**

1. What is the method of producing goods in large quantities called?
2. Why have most companies introduced automation in their factories?
3. How are technical drawings made nowadays?
4. What advantages does the CAD system have?
5. What does a CAM system control?
6. In what way is the CAM system flexible?
7. What electronic equipment do modern robots have?
8. Why are they equipped with these things?

c The most advanced systems link together CAD, CAM, robots and other automated machinery. One such system is called Computer Integrated Manufacturing (CIM). It is shown on the right.
Describe in your own words how it works.

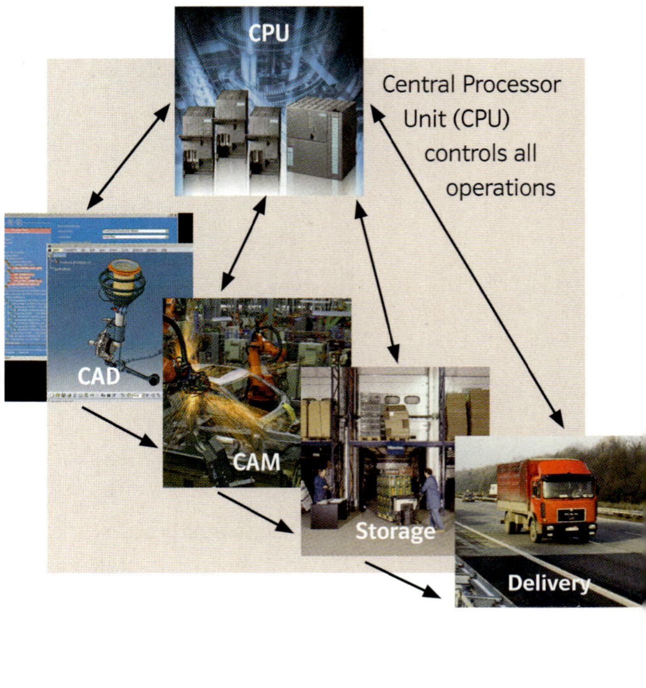

Freeway

S. 71 ✓, 72 ✓

S. 67

Focus

S. 166, 167

technical expert

S. 60/61

Talk Tech

S. 78 ✓/79 ✓/ 180 ✓/81 ✓/182/183

advertisment

describing products

2 Describing a production system

Match the words in the box with the parts of the car.

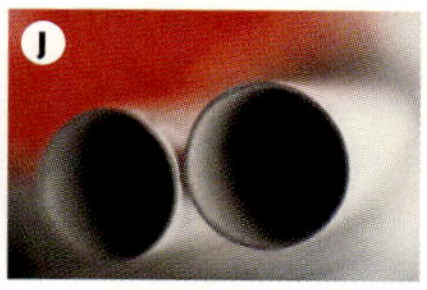

1. engine
2. doors
3. headlights
4. wheels
5. dashboard
6. windscreen
7. bumper
8. exhaust pipe
9. car body
10. steering wheel

b In the diagram below you can see how a car is made in a modern car factory. **Describe the production system. Use the vocabulary in the box.**

Example:
First lorries deliver the different parts of the car to the factory.

lorries workers machines robots	assemble, add, test, dip, weld, dry, fit, deliver, drive away, put on

1

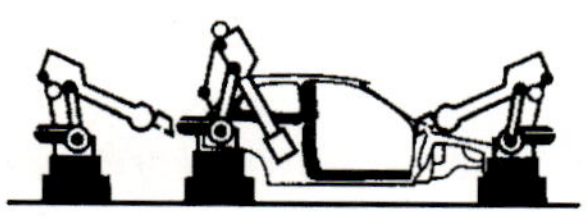

2

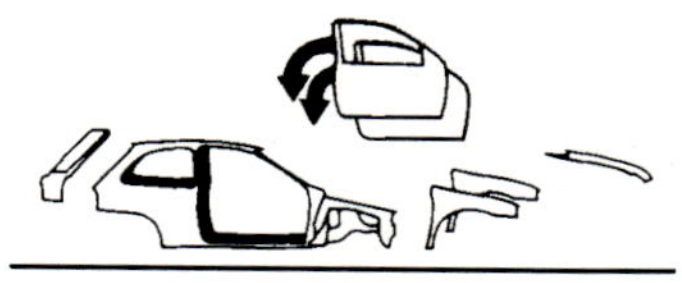

3

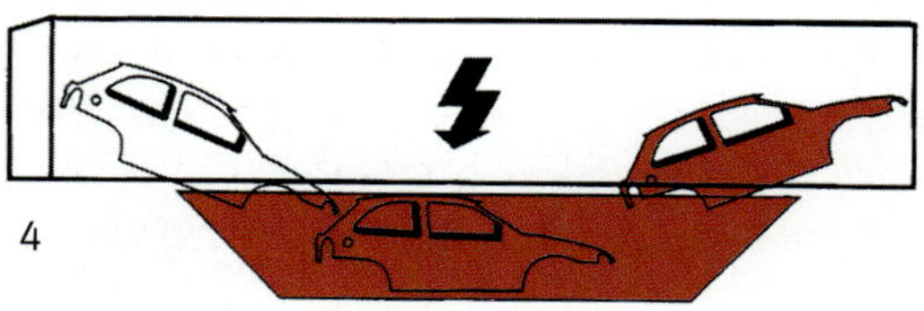

4

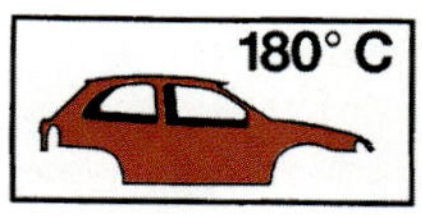

5

6

7

8

9

10

1 How is a car produced?

■ Describe the production process on page 67 in the passive voice.
Add "by" if necessary.

Example:
The parts are delivered by lorries. The body is ...

2 How should production be improved?

In order to succeed on the international market a modern car has to be produced in only a few hours today. How can this be done?
Here are some suggestions the workers have made.

■ Complete the sentences in the passive voice. *Must, should, can* or *could* should be used at least once.

Example:
1. The workers should be trained more carefully.

1. workers / train more carefully
2. production process / organize better
3. cars / assemble / by groups of workers
4. important parts / make / by the workers
5. other parts / deliver "just-in-time"
6. better working conditions / introduce
7. a new kind of fast-drying paint / use
8. workers / give more responsibility
9. workers' suggestions / follow / by the management

3 Production in the past and today

Many years ago motor companies used only a few machines. So the workers did most of the hard and dirty work. After Henry Ford had invented the production line, motor companies started mass production. In the last few decades companies have improved the production process. Nowadays they are still developing new production methods. If they did not do this, they would sell fewer cars, because of growing international competition. So companies have bought a lot of expensive machines. Today production lines transport the car bodies and robots do most of the work. However, machines cannot replace all workers and companies will still need qualified staff in the future.

■ Describe the development of production in the passive voice.

Example:
Many years ago only a few machines were used by motor companies.
So most of ...

4 A new job

Wesley Harding has applied for a job at a big motor company. This is what happened to him:

1. Before his interview/Wesley/a letter/send/the personnel manager
2. On the day of his interview/Mr Harding/questions/ask/about his qualifications/the personnel manager
3. While he was waiting/the applicant/a cup of tea/serve/the tea-lady
4. Then/he/the new machines/show/the head technician
5. Afterwards/Wesley/lunch ticket/give/a secretary
6. In the afternoon/he/result/tell/the personnel manager
7. Wesley/job/give/the personnel manager
8. Before he left/Mr Harding/£20 travel expenses/pay

■ **Make sentences both in the active and passive voice.**

Example:

1. Before his interview the personnel manager sent a letter to Wesley.
   … Wesley was sent a letter by the personnel manager.

5 Translation **Advanced**

■ **Translate the following text. Watch out for the tenses.**

Die meisten Automobilfirmen wurden vor vielen Jahren gegründet. Zu jener Zeit wurden viele Arbeiter benötigt, um Autos zu produzieren. Es konnten pro Tag natürlich weniger Fahrzeuge als heute hergestellt werden. Aber nachdem neue Technologien eingeführt worden waren, wurden mehr und mehr Arbeiter entlassen (dismiss). Seit dem Beginn der Massenproduktion ist die Produktivität erhöht worden. Heute wird die Produktion von Robotern ausgeführt und von Computern kontrolliert. Der Prozess der Automatisierung kann nicht angehalten werden. In der Fabrik der Zukunft werden wahrscheinlich viele Produkte ohne Arbeitskräfte hergestellt werden. Allerdings werden auch dann qualifizierte (skilled) Arbeiter benötigt werden, weil nicht alle Tätigkeiten von Maschinen ausgeführt werden können. Zur Zeit werden Arbeitnehmer von verschiedenen Autoherstellern zu ihren Ideen befragt, wie die Produktion verbessert werden kann. Neue Technologien können auch zusammen mit den Gewerkschaften entwickelt und eingeführt werden. Aber natürlich weiß heute niemand genau, welche Autos in fünfzig Jahren produziert werden und wie der Produktionsprozess dann aussehen wird.

1 Team work works

Automation, computer-controlled robots, in recent years we have all seen pictures of modern production lines which are dominated by machines and where no humans can be seen. However, in one German electronics firm in Cardiff, Wales, humans are making a comeback. Personnel manager Martin Wibberly told me "Nowadays we think of our workers not only as a cost factor, but as capital and so we invest in them." And the company invests in them because their flexibility is needed in today's market. Martin added "Flexibility means a faster and more customer-orientated service. Of course, we still use robots but the production lines of the past were too inflexible and complex and therefore too expensive to serve today's market. And so these methods have been replaced by much smaller and individual production units."

It's an arrangement which suits the workers equally well. Peter Brown, who was one of the 500 chosen from 25,000 applicants for jobs at the firm, explained the advantages of the new system for the workers. "As you know, past production methods were all based on mass production where each worker on the production line performed one small monotonous task. Now however, the work is done in groups which are completely responsible for one individual step in the production process. This arrangement is also more flexible for us because now we can arrange our own working schedules."

Today there are also opportunities for workers to train and get further qualifications so that they can do different jobs when this becomes necessary. This too, means that the work is much less repetitive and boring than in the past. The groups also meet regularly to discuss problems in the production process, to suggest solutions and to help to put them into practice.

Peter Brown points out that when the workers have a say in the organization of their work, communication between workers and management is improved. He also enjoys the feeling of responsibility not only for his own work, but also for the success of the whole process.

These changes have been supported by the unions. They too, recognize the opportunities for personal development which workers will be given under this system. Lean management and productivity in the Japanese style can only be effective, they say, when all the employees have more say in the decision-making process.

387 words

a Name two methods of production mentioned in the text.

b Find three groups of people who support the new method.

c List the disadvantages of the old system.

d List the advantages of the new system.

e Give the meaning of the following expressions from the text.

1. customer-orientated service (line 13)
2. mass production (line 26)
3. working schedules (line 33)
4. further qualifications (line 35)
5. lean management (line 52)
6. decision-making process (line 55)

f Write a summary of the text.

g Translate lines 34 to 48.

2 Describing a process

a A large electronics firm in Germany has developed a new DVD recorder. They need the recording instructions in English. The production manager has given you these pictures.

Write the instructions in English for the export model.

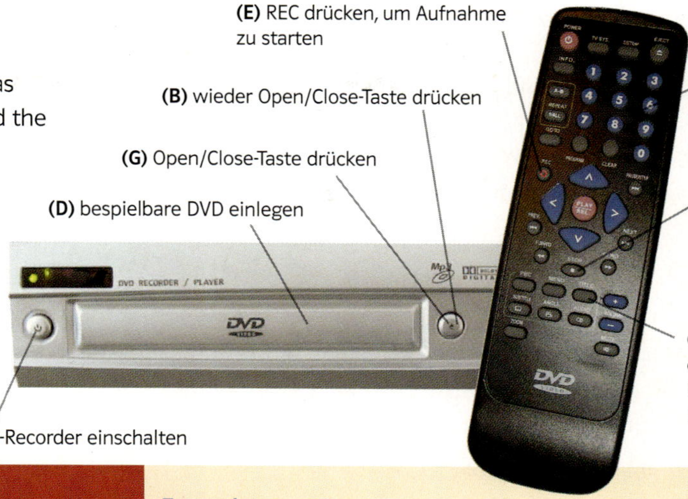

Follow these steps

1. Put the instructions in the right order.

Example:

(F) DVD-Recorder einschalten
(G) Open/Close-Taste drücken
(D) bespielbare DVD einlegen
...

2. Translate the instructions using the words in the box below. If you need more help, use your dictionary.

Switch DVD on – Press OPEN/CLOSE button – Insert ...

einschalten:	to switch on
drücken:	to press
einlegen:	to insert
auswählen:	to select
aufnehmen:	to record

3. Use infinitives, modals or passive forms.

First, the DVD recorder must be switched on. Then ...

4. Ask your partner to test the instructions at home (if possible).

b Here is a new cassette recorder which has been produced by the same firm.
Write the operating instructions for a recording.

Info-Box: How to describe a process

Remember:

1. **Put your information in the right order.**
2. **Use your dictionary to find exact meanings.**
3. **Write precise instructions. You can use infinitives, modals or passive forms.**
4. **Test out the instructions on the equipment.**

1 Sales talk

a Listen and find out the following things:

1. Who is talking ?
2. Where are they talking ?
3. What are they talking about ?

b Listen again and find out about the following:

1. speed
2. paper
3. print quality (resolution)
4. installation
5. service
6. price

2 Installing equipment

You have just bought the new Pineapple ink jet printer.
Unfortunately the instructions in the handbook are only in English. You also discover that these instructions are completely mixed up.

a First find the correct headlines for the instructions. Choose from the headlines in the box below.

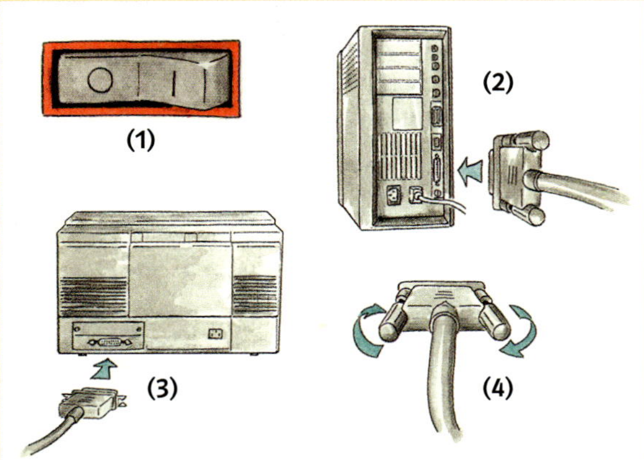

Instruction 1.
- Tighten the screws of the interface connector.
- Connect the other end of the interface cable to your printer.
- Make sure that both the computer and the printer are turned off.
- Connect one end of the interface cable to the computer.

Headlines:
- Printing
- Connecting the printer
- Functions
- Installing the ink cartridge
- Fault finding

Instruction 2.
- Put the printer cover back on.
- Press the cartridge down until it snaps into position.
- Take off the safety tape on the ink cartridge.
- Remove the printer cover.

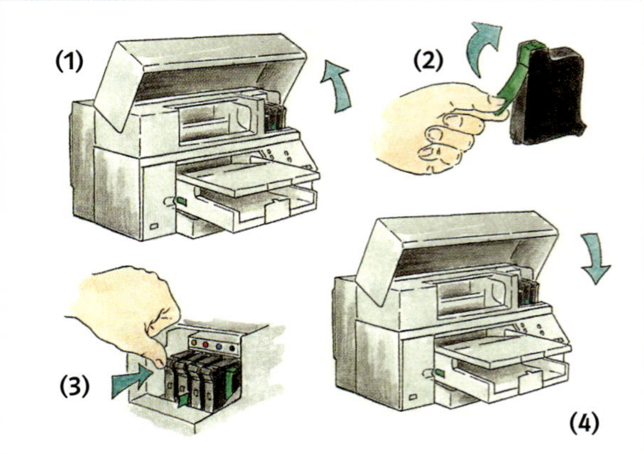

b Now put the instructions into the correct order. The drawings will help you.

c Translate the instructions into German.

d You still need help to install the ink jet printer. The supplier of the printer comes to your house and shows you how to do this.
Work out the conversation you might have with this computer expert. Play these roles.

Changes in British Industry

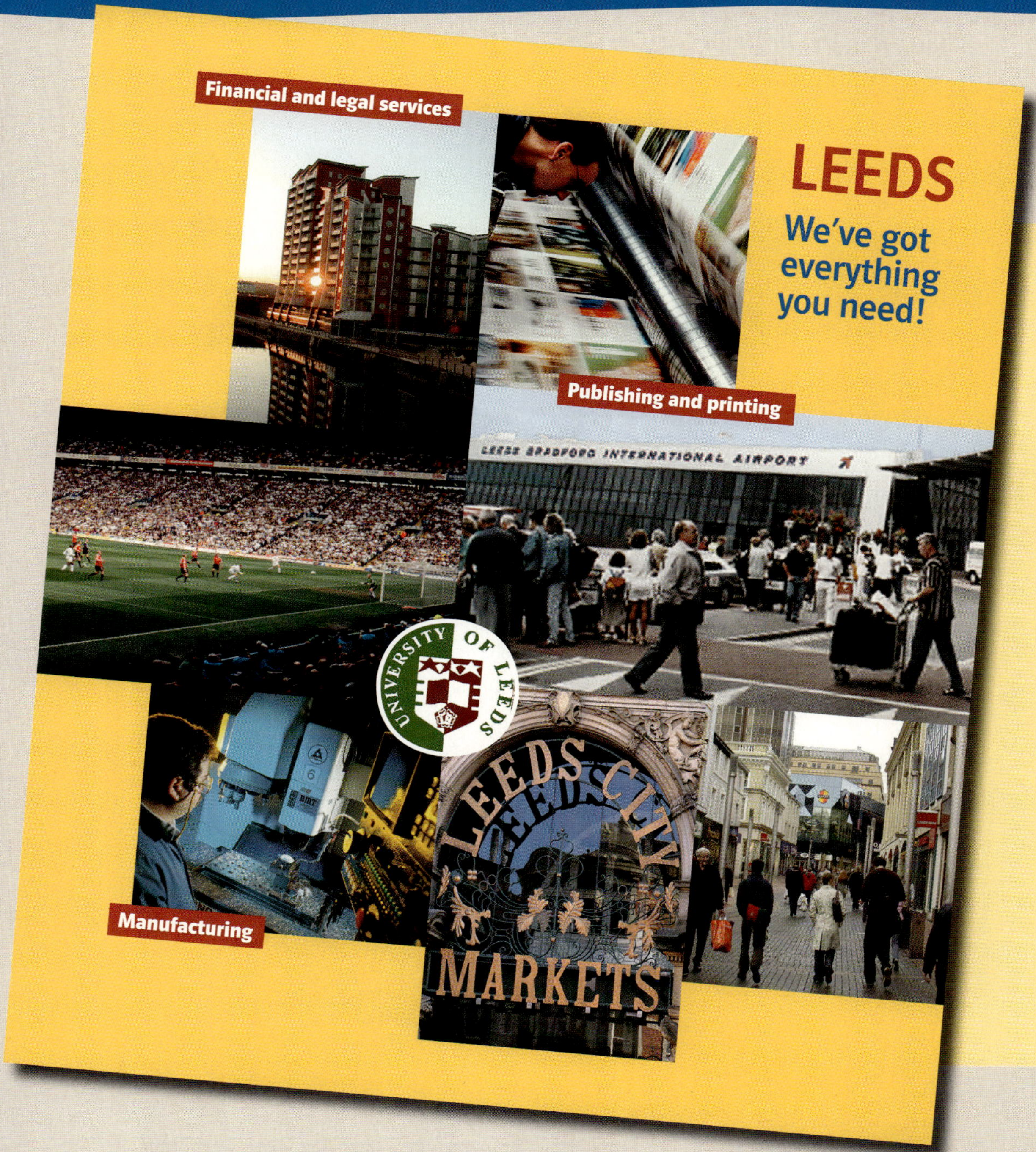

a What do the photos tell us about Leeds?

b Who do you think wrote the brochure and why?

c Compare your town / region with Leeds.

d Imagine you have to write a brochure in English for your town / region. Think of a slogan. What photos would you use? Give reasons for your choice.

1 Changes in the North of England

Look at the industrial map of Great Britain in the 1960s and the newspaper headlines on the left. Now answer the following questions.

1. What kind of industries could you find in the Leeds/Newcastle area in the sixties?
2. What has happened to them?
3. What has happened to the workers?
4. Why do you think has there been such a development?

2 Newcastle takes off

Newcastle Business Park, the £100 million development of a former industrial area, was officially opened some years ago. The success of the Newcastle scheme was that 95 per cent of the 67,000 square metres of office buildings had been let immediately, mostly to computer companies and light industries. The management of the business park gave financial help to many of these new companies and provided them with the latest communications technology.

The business park is ideally located, as the area has the advantage of excellent road, rail and air links. Today more than 4,000 people work on the 1.5km site.

Decades ago the site was occupied by a large factory which produced machinery and arms. When the factory closed down, nobody was interested in taking over the site. However, unemployment figures were so high in Newcastle some years ago that the Newcastle Development Corporation decided to invest £100 million in this site. John Scally, the managing director, proudly said: "Newcastle Business Park is an outstanding example of what cities can achieve."

Newcastle City Council has also recently started an exciting £10 million partnership with Digitalbrain Plc to become Europe's first Digital City. This programme will provide businesses and citizens with a range of interactive electronic services and information to improve the region's economic growth. Newcastle upon Tyne is now one of the most wired cities in terms of broadband network coverage. This allows companies and private individuals access to large-scale data processing. Moreover, Newcastle City Council is creating a Centre for Entrepreneurship and Innovation (CEI) in the middle of the Newcastle Business Park. These buildings will provide office space for new companies in the area of digital technology and will create over 400 jobs. As you can see Newcastle is taking off!

295 words

a What has been done in the Newcastle area in the last few years?

b Answer the following questions.
Some answers cannot just be taken directly from the text.

1. What does a business park look like?
2. What was produced on the site of the new business park years ago?
3. Who (do you think) has given the money for the project to the Newcastle Development Corporation?
4. Why (do you think) has so much money been invested?
5. Why (do you think) are so many companies interested in the new office buildings?
6. Why (do you think) were they not interested in the old site?
7. In what way (do you think) is Newcastle becoming Europe´s first Digital City?
8. Why (do you think) is Newcastle creating the CEI?

c Now think of the town/area where you live. Describe a similar project that you have seen or heard of.

1 A Canadian student at Newcastle college

Craig Russell, a Canadian student from Vancouver, is planning to go on a year's exchange visit to Newcastle College in the north of England. He has never been to Europe before. So yesterday he asked Calvin, another student who was on last year's exchange, for some help. Here are some of Craig's questions and the answers he received.

a Match the answers to the questions.

1. Who should I contact first at the college?
2. When should I write to my host family?
3. How should I travel from London Airport to Newcastle?
4. Where can I reserve a seat for that journey?
5. What should I give my host family as a present?
6. How much luggage should I take?
7. What should I wear at college?
8. Where can I find a good bank in Newcastle?

a) at a travel agent's
b) a month before you arrive
c) opposite the college
d) Ms Keegan, the head of department
e) not too much
f) by coach
g) casual clothes
h) something typically Canadian

b Now report Calvin's answers like this:

Example:
Calvin told him / advised him to contact Ms Keegan at the college.

c Craig has been at Newcastle College for one month now. This is part of Craig's letter home. As you can read the teachers **expect him to work** hard during his stay in Newcastle and sometimes they **make him do** some extra work. Complete the letter with the verbs in the box.

let • expect • make • want • ask

Life at college is great. I am the second Canadian student to study here for a whole year. I get on well with the teachers here but they **(1)** me work quite hard. They **(2)** me to talk to the English students more about the Canadian college system. I didn't realize it was so complicated. In fact, the teachers **(3)** me to know a lot more than I actually do. Last week, for example, they **(4)** to show my photos of our college in Vancouver in order to give the English students a better idea of our work at Canadian colleges. But it is not all hard work. And if I want to do some sight-seeing at the weekend, the teachers **(5)** me go early on Fridays. By the way I have met …

2 Finding out useful information

a On the next page you can see parts of the students' Internet notice board at Newcastle College. Match the following sentences with the notices on the Internet notice board.

Example: Sentence 1 goes with notice D.

b Complete the sentences below. Use the -ing form of a suitable verb.

Example: I'm interested in knowing more about France so I'll go to the French Club.

1. I'm interested in … more about France.
2. I'm tired of … with my parents and I need somewhere to live.
3. Students who are afraid of … their final tests should ring 347168.
4. … is dangerous for your health.
5. Craig is going to the club because he is good at … basketball.
6. We'll leave the car at home … and … is stupid.
7. I'm looking forward to … you on Friday evening.
8. If Gary takes me in his car, I won't have any more problems in … to college on time.

3 Talking about friends

Craig has met a lot of new friends at Newcastle College.
Here are some remarks they have made about themselves.

Sharon: "I play more hockey than volleyball at the Sports Club."
Kevin: "I go to the cinema every week."
Emma: "I don't smoke any more."
Bill and Ben: "We don't eat in the college canteen. The food's not very good there."
Sam: "That's one thing I hate – my alarm clock."
Mary Ann: "I'm very interested in foreign languages."
Gary and Sue: "We did the same test last year. "
Lisa: "I try not to be late for college. I take the early bus."
Mike: "I spend all my free time at the Sports Club."
Tina: "This book is very useful. I think you should read it."

Make statements about Craig's friends.
The verbs in the box will help you.

like • hate • prefer • remember • stop • enjoy • avoid • suggest • dislike • love

Example: Sharon prefers playing hockey to volleyball at the Sports Club.

4 Translation **Advanced**

Imagine you were at Newcastle College on an exchange visit and that you had to write a report in English about your experiences there.

Translate these notes into English.

1. Meine Gastfamilie ist sehr nett. Ich wohne gern bei ihr.
2. Aber zuerst hatte ich Schwierigkeiten, ihr Englisch zu verstehen.
3. Am Anfang hatte ich Angst, Fehler zu machen.
4. Die englischen Schüler im College sind sehr interessiert, mehr über Deutschland zu hören.
5. Im College lassen die Lehrer mich ziemlich hart arbeiten.
6. Sie erwarten, dass ich viel über das deutsche Schulsystem weiß.

'Made in Britain' loses appeal

1 When he invented the bagless vacuum cleaner – among other things -, Britons looked upon James Dyson as just one more in their long tradition of inventors. When Dyson, who is now 54, turned his ideas into a private business with sales of $315 million in 2000 and an estimated value of $700 million, the inventor became an icon of innovative Britain. His name has become a brand name and is printed on all of the 8 million Dyson bagless vacuum cleaners his company has sold since 1993.

2 But things are changing. Although he had promised that he would never leave his home country, Dyson shifted production of all his vacuum cleaners to the cheaper labour markets of Asia last year. Moreover, he has just announced that he is planning to close an assembly line for washing machines in Britain and open a new plant in Malaysia. The move will eliminate hundreds of jobs at his relatively new plant in Great Britain, saving labour and other costs as Dyson prepares to introduce his products into the U.S. market this summer. "I have put £40 million of my money into this business to try and make manufacturing work here," Dyson said. "But I have to regard the law of economics. If we want to survive as a business, we have to go where manufacturing is economical. It is only in this way that we can keep more than 1,200 hi-tech jobs in research, design and engineering.

3 In the past three decades, manufacturers have seen their share of Britain's annual output go down by 50 percent while services such as banking or restaurants have prospered in a new kind of consumer economy that has given Britons jobs, low interest rates and huge spending power. "The country as a whole is in a two-speed economy where the consumer economy is roaring away and manufacturing is in a deep recession," said Eric Britton, an economist. "It is typical of developed economies that the manufacturing sector tends to go down. We must accept that manufacturing is a global industry."

4 But while much of Continental Europe is struggling with unemployment, the British unemployment rate is down to 3.2 percent. Britain, once called by Napoleon a nation of shopkeepers, has become a nation of shoppers: Consumer spending accounts for 60 percent of the economy.

5 Contrast that with the old-fashioned business of producing things to sell: Last year, 150,000 jobs were lost in the manufacturing sector as its share of national economic activity fell below 20 percent. By contrast, service industries such as call centres added almost 230,000 jobs. But does it matter? As long as consumers continue to spend the way they are doing – stores recorded an average increase of 7 percent for the Christmas holidays – growth will continue.

468 words

a Each of the following sentences represents a paragraph of the text. Put the sentences in the order of the paragraphs.

1. Dyson has to close down factories in Britain in order to save costs.
2. Compared to other European countries the economic situation in Britain is quite good.
3. Many jobs in manufacturing are replaced by jobs in the service industries.
4. Inventor James Dyson has become a well-known and successful businessman.
5. While the manufacturing sector is decreasing in Britain consumption is rising.

b Answer the following questions in complete sentences.

1. What is James Dyson famous for?
2. Why is he closing production plants in Britain?
3. What has happened to the manufacturing industries in developed countries.
4. What is the job situation like in Britain?
5. In which fields can jobs be kept in Britain?

c "Workers should accept worse conditions (wage, working hours, working conditions, no strikes) in order to save their jobs or to create new jobs." Do you agree? Give your comment.

Write a comment.

Follow these steps		Example:
1.	**Write a general introduction to the topic, e.g. give reasons why the topic is of special importance today.**	Today the unemployment rate in many countries is high. So a lot of workers fear for their jobs, and the jobless are desperately looking for new jobs. Many of them are willing to accept lower pay and worse working conditions than before.
2.	**Collect ideas for arguments for and against the statement above and put them in a suitable order.**	Arguments for: - lower costs for companies - more competitive on international markets Arguments against: - lower standard of living
3.	**Then make complete sentences and decide which phrases may be suitable** **e.g.: Firstly/In addition/On the one hand/On the other hand/In contrast, …**	On the one hand this brings many advantages to the companies. Firstly …
4.	**Draw conclusions from the arguments above and state your point of view. Use these phrases:** **In conclusion/As a result/In my opinion/In my view/It is for this reason that …**	In conclusion I would say that workers should/should not …

d Translate lines 11 to 28.

Info-Box: How to write a comment

Remember:

1. **Write a general introduction to the topic.**
2. **Collect ideas for arguments for and against the statement and put them in a suitable order.**
3. **Choose expressions to present your arguments and write complete sentences.**
4. **Draw your conclusions from the arguments given and state your personal point of view.**

Looking for a new location

Imagine you work for Samsons Ltd, an international manufacturer of high class racing bicycles. They are very interested in finding a new site for a new factory and offices in England. Your company has looked at a lot of promotion literature for development areas in Britain.

Extract 1

England's Largest County is Getting Bigger!

Large investments by private developers in the North's most attractive county allow us a wonderful opportunity to invite you to join our economic success. Fifteen new office and business parks are being developed both in town centres and out of town. Offices with up to 25,000 square feet will soon be available, while offices up to 35,000 square feet can be built to your own requirements.
North Yorkshire has a growing population, top schools, colleges and universities, good housing and shopping facilities, excellent road and rail links (only 2 hours from London) and highly developed business contacts.

Write or phone today for details of current opportunities.

The Economic Development Centre,
North Yorkshire
County Council, County Hall,
Northallerton, North Yorkshire DL7 8AD, England.
Telephone 0609 780780.
http://www.edcinfo.co.uk

Extract 2

Where on earth should we relocate?
The answer is easy: to the Black Country.

We're right in the centre of England which means there is direct access to London and the Channel Tunnel via the national motorway network. What else does the location offer? Plenty of skilled labour (after all we have represented the heart of Britain's manufacturing base for more than 200 years). Thousands of square metres of land for offices and factories. The buildings which have already been built are of the highest standard and we have not forgotten to build large car parks! In addition, we can provide all the back-up you need; from planning permission to expert help, even financial assistance. The Black Country – your future in Europe has never looked brighter.

For more information,
contact Linda Clement on **44 21 511 2000,
or write to her – Black Country Development Corporation, Black Country House, Rounds Green Road, Odbury, West Midlands B69 2DG, England.
http://www.bcde.gov.uk
email: clement@blackcountry.co.uk

a Read the two extracts above and do the following tasks.

1. Find out where exactly the two areas are.
2. Make a list of the key points in both of the extracts.
3. Find out from your key points
   - which points are mentioned in both texts.
   - which are only in one of the texts.
4. Give a short report about the two areas.
5. Discuss your results in class. Which area do you think is better for your company? Give reasons.
6. Write a letter to both areas stating your situation, and ask for more information.

b You are interested in finding out more about the two regions. Two colleagues have visited the areas. Listen to their reports and find out from their phone calls:

1. which towns they have visited.
2. where these towns are.

c Listen again and find out more about the following:

1. transport links
2. industry
3. price of land
4. housing
5. job situation
6. general impression

d Which town would you now choose for a new site for your company?
Work in groups and then write a report and give reasons for your decision.

Immigration in the United States

Two hundred years ago the U.S. had a population of about 4 million people. Today the population has risen to more than 280 million.

Immigrants – Where do they come from?

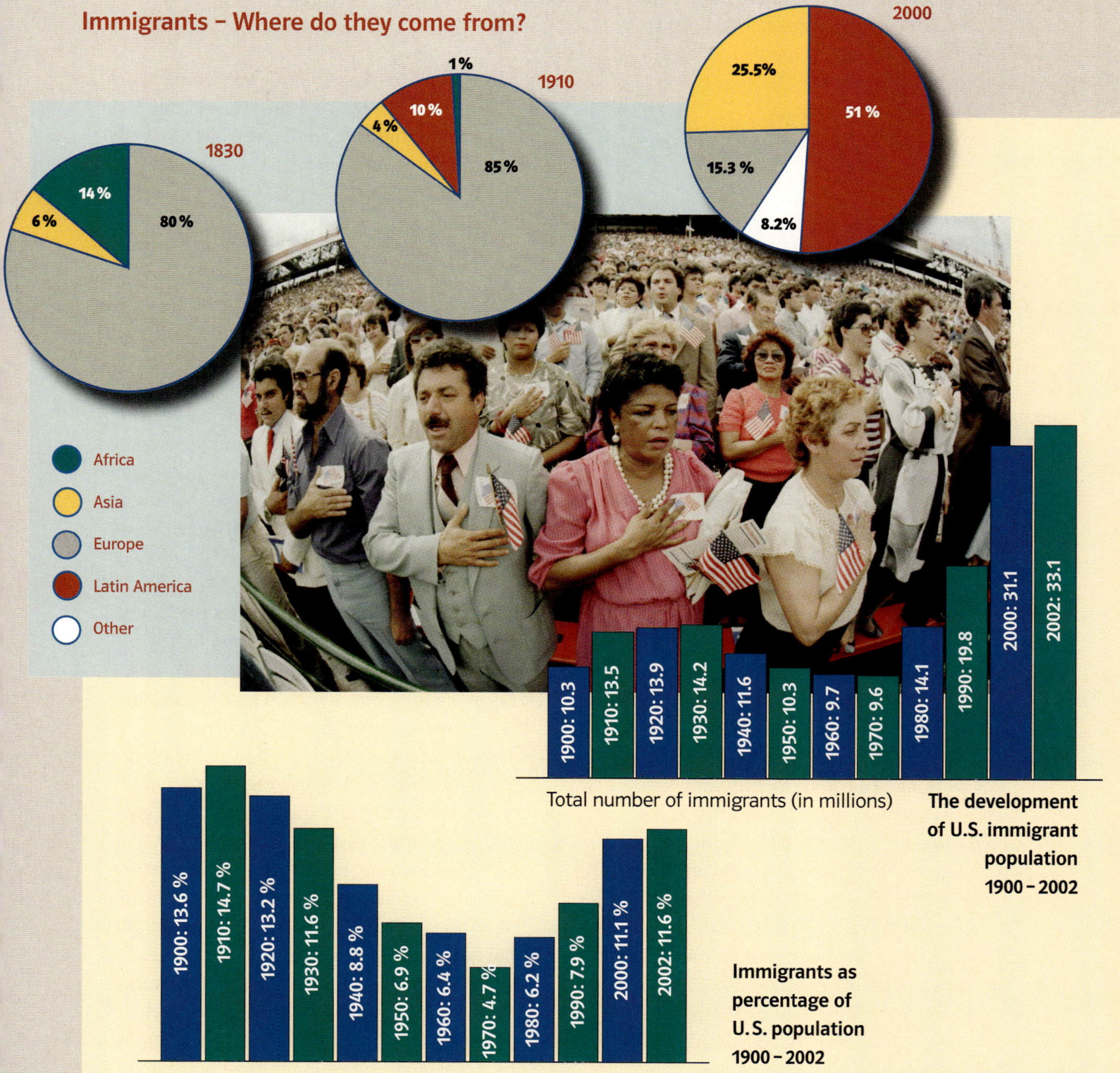

a Describe the development of immigration in the United States.

b Where have the different groups of immigrants come from over the years?

c What reasons have they had for coming? What do you think?

1 Two faces of America

a What do you think the title "Two faces of America" means?

b Find out the following information from the texts below:

1. where the people came from originally
2. when they arrived in the U.S.
3. where they live now
4. what they do in America
5. why they emigrated

Marek and Lisa Leschinski came to Chicago from Poland with their two sons seven years ago. They were fleeing from nothing more than a middle-class existence. Marek, now 32, was a technical draughtsman in Warsaw. Lisa was a receptionist in a hotel. "We had a small apartment, jobs, a car and everything we basically needed," says Marek, "but there were not really any chances for more." After three years of trying to get a visa to the States, they finally arrived in Chicago.

Marek first had to take a job as a machine operator with wages of $310 a week, which is not much when you have a family to support. However, a year later he had more than doubled his salary in another job. Lisa was working in a cookie factory, and they were saving most of the money that came in. Soon they had saved enough for a down payment on 'The European Cafe' on North Lincoln Street, above which they live in a 5-room apartment. A year ago they bought the 'Town Bakery' in a fashionable suburb of Chicago. Marek says they earn $ 1500 a week now. They have to work 16 hours a day, doing everything from baking to making the deliveries, but they are happier than they have ever been before. As Marek puts it, "In this country you have the chance – if you work more, you can earn more." **236 words**

Growing up in Vietnam, June Yen learnt to be an outsider. She is a so-called Amerasian. Her father is an American soldier she has never met. Her Vietnamese mother spent years explaining to her why she looked different to her friends. When she came to the States years ago she thought she would finally become integrated. She was wrong. "In Vietnam they called me American," June explained, "here, they don't know who I am and I still don't really know where I belong." After arriving in America she went to school in Atlanta. There she was put into a class with children half her age because her English was not good enough. This made her feel like an outsider from the beginning. To make matters worse one teacher there, who had fought in the Vietnam war even told her, "Why don't you go back to your own country?" This made June very sceptical and even afraid of her American teachers. June still cannot speak or write English very well and she often feels like a second-class person. After leaving school without any real qualifications she got a job in a Chinese restaurant. The pay was bad and the hours long. Four years ago she met her Vietnamese husband. A year later they moved to Memphis where her husband found a new job. June has given up work and now looks after their two children at home. As June says "Emigration to a new country takes a lot of courage and determination. Arriving in a foreign country with practically no possessions and not being able to speak the language can be a very frightening experience. I only hope my children will have a better start in life than I did." **293 words**

c **Find out from the texts why the following statements are true.**

1. Marek and Lisa were not poor immigrants.
2. It took 3 years before they could emigrate to America.
3. They started their own business in America.
4. The Leschinskis have found happiness in America.
5. June Yen had identity problems in Vietnam.
6. She had problems at the school in Atlanta.
7. She still feels like a second-class citizen.
8. Her children will hopefully have a happier childhood.

d **Discuss the following quotations from the texts.**

1. "In America you have the chance - if you work more, you can earn more."
2. "Emigration takes courage and determination."

2 American immigrants: who are they and where do they go?

a **Link the five statements below with the states shown in the graph.**

1. During the riots in L.A. in 1992 tension grew between the black community and the new immigrants. 2,000 Asian owned businesses were looted or damaged by fire.
2. More Poles live in Chicago than in any other city in the world except Warsaw. The city still attracts new Polish immigrants.
3. Ellis Island closed as a port of entry in 1954, but this city in the east still attracts more immigrants than any other city in America.
4. People from Latin America are still coming in great numbers to Miami.
5. Along its Mexican border, which is 1,200 miles long, about 400,000 illegal immigrants are arrested every year.

b **Find out from the graphs the differences and similarities between the eastern and the western states.**

c **What else do the figures tell us?**

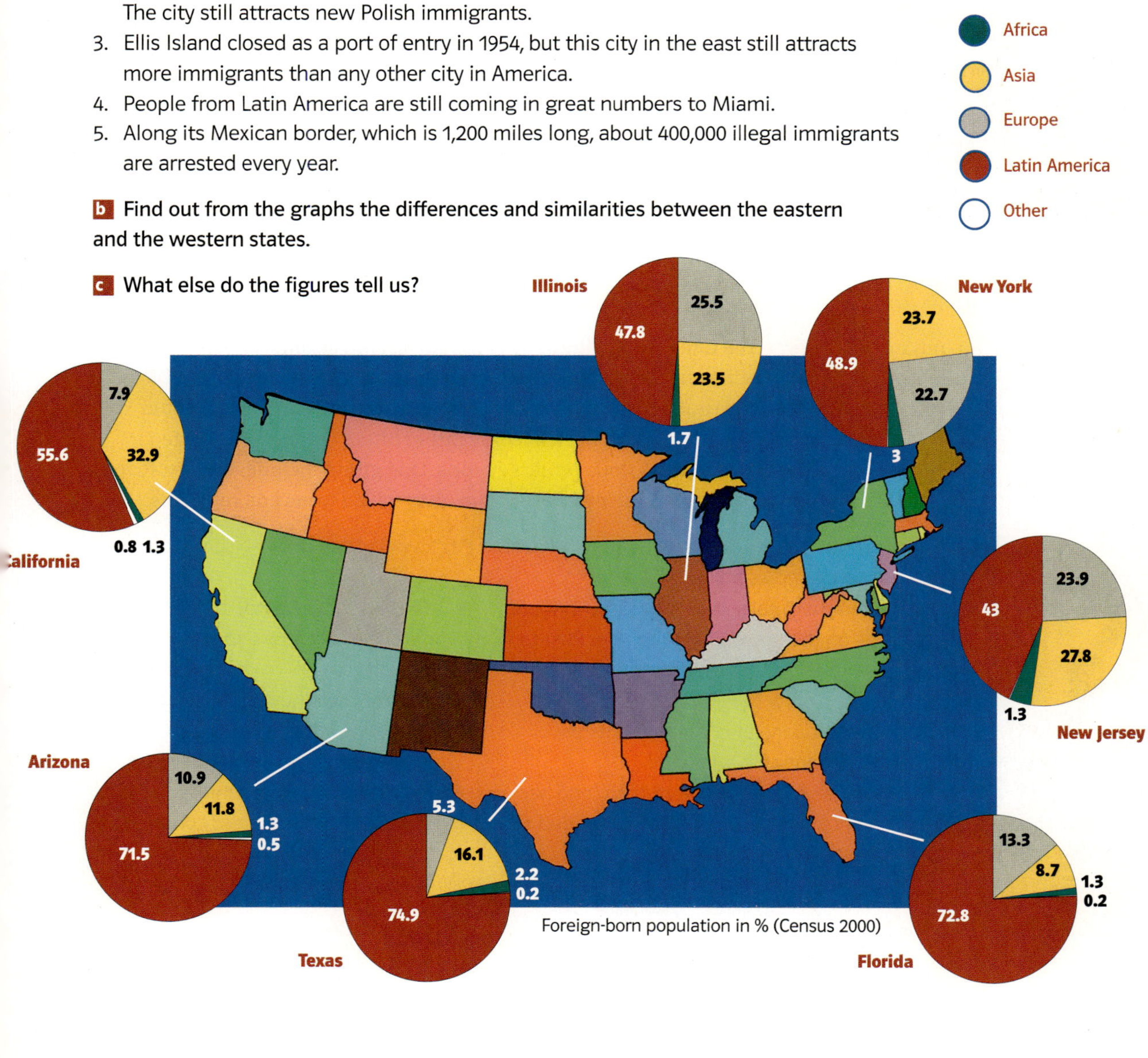

Foreign-born population in % (Census 2000)

1 California and the Chinese

The Californian population is a mixture of immigrants from many different nations **which includes** lots of Chinese. Most foreigners know the part of San Francisco **which is called** Chinatown.

Today Chinatown has not only a lot of fine restaurants **which have been built** for the tourists, but it is also a place where some Chinese people do not have enough to eat. In some areas there are families with four or five children **who live** in a single room. And so you find many families **who use** the same kitchen and bathroom. Today Chinatown is not only a tourist attraction but also a place with a lot of poverty. Meanwhile many young Chinese have left Chinatown for other places **which offer** better jobs. But most of the new immigrants **who come** from China cannot move because they do not know the language **which is spoken** outside the Chinese community. The older Chinese people **who have been** there for decades do not want to move to another place. They hold on to their old ways, and they do not often ask for help from outside the community.

■ Replace the words in bold type by a present participle (ing-form) or past participle (third form of the verb).

Example:
The Californian population is a mixture of immigrants from many different nations **including** lots of Chinese. Most foreigners know the part of San Francisco **called** Chinatown. Today Chinatown …

2 Native Americans in California

The American Indian was the first true Californian. The style of Indian art, music and clothing is part of the whole West, including California. But like other minorities, the Indians have had trouble in Californian society. In their case the problem is land. **As they believe** that they have been treated unfairly by the American politicians, many of them have become rather bitter. Others have started to make their voices heard in various protest actions. One group includes the Pitt River Indians. **Before they started** their protest action, they talked to several reporters.

■ Now put the following statements about the Pitt River Indians into the same form as in the examples below.

1. When they realized how much gold there was on their land, the whites decided to take it away from the Indians.
2. As they knew the Indians had been treated badly, later governments gave them money for their land.
3. Because the Indians want their land back and not the money, they have decided to start a protest action.
4. After they have made their protest here, they are going to fly to Washington.
5. As they know how difficult it is to get a meeting with the President, they are prepared to wait as long as necessary.
6. Before they leave Washington the Indians will give a press conference.

Example:
Believing that they have been treated unfairly … **Before starting** their protest action …

3 A Mexican immigrant

A Mexican immigrant talks about his life in the U.S. during a break in his work.

■ Infinitive or ing-form? Fill in the correct form of the verb.

Example:
1. I'll never forget leaving my small Mexican hometown.
2. My mother told me:"Don't forget to write to us …

I'll never forget **1. (leave)** my small Mexican hometown. My mother told me:"Don't forget **2. (write)** to us when you're there. " I remember **3. (cross)** the border by night. I was afraid of **4. (be arrested)**, but I wanted **5. (change)** my life completely, so I decided **6. (take)** the chance. In the first town I stopped **7. (get)** some food. Inside the restaurant there was a sign **8. (say)** that they needed help in the kitchen. I remember **9. (ask)** the boss what he was prepared **10. (pay)**. It wasn't much, but **11. (need)** the money urgently I accepted it.
They made me **12. (do)** the dirty jobs. First I enjoyed **13. (work)** there because the other workers were friendly. But after **14. (stay)** there for six months I moved here to San Diego. I hope **15. (find)** my own apartment soon. I would like **16. (stay)** here forever. That's why I must remember **17. (phone)** the Housing Officer tomorrow. I used **18. (be)** a so-called illegal immigrant, but a couple of years ago I got a green card. Now I'm able **19. (live)** and work wherever I like.
– Sorry, my break is over. I must stop **20. (talk)** now. I have **21. (get)** back to work again.

4 A German immigrant **Advanced**

Ursula Zimmermann from Germany emigrated to America twenty years ago, but she has always kept in contact with her friends in Germany.

■ Translate the following part of one of her e-mails.

Vor einigen Monaten sah ich öfters drei junge Leute im Supermarkt. Ich erinnere mich, dass ich Herrn Cheng, den Leiter des Supermarktes, gefragt hatte, wer sie waren. Er erzählte mir, dass sie deutsche Studenten waren, die eine Reise von Kalifornien nach New York planten. Ich erinnere mich, dass ich sie eines Abends traf. Sie reparierten gerade ihr altes Auto. Ich hielt an, um mit ihnen zu reden. Es machte mir richtig Spaß, wieder Deutsch zu sprechen. Die jungen Leute freuten sich darauf, die USA zu durchqueren und die vielen Sehenswürdigkeiten zu sehen. Ich bot ihnen an, bei der Planung der Reise zu helfen.

Immigration and welfare

Together with corn and cars, immigration has been one of the main factors of American economic growth. The traditional theory is simple: new workers increase the supply of goods and services with their labor and increase the demand for other goods and services by spending their wages. A circle of growth begins as a growing number of workers create a richer society for each other. Two hundred years of U.S. history seem to confirm this theory. But there is a feeling today that immigration is not so good for the economy any more.

The polls show that many U.S. citizens are worried that more immigrants will take jobs away from native-born workers. This can be true in times of high unemployment. In California, for example, a recent survey shows that the population explosion between 1990 and 2000 was the result of a massive inflow of immigrants. With a jobless rate close to seven per cent (2003) many native-born Americans are actually leaving to find work in other states. So we must see both sides: "The short-term costs of immigration are much higher today," says Michael Boston, one of the leading American economists, "but in the long run, immigrants are still great news for our economy. Normally, new jobs are created through the immigrants' own work. The immigrants' spending creates a demand for houses, food, etc. and their employers invest their growing profits in new machinery and jobs. That's how America got rich."

However, in the last few decades things have changed. In the early days of immigration public education and some public health programs were the only services for those coming to New York or other Northeastern cities. One third of the new immigrants soon found out that America did not offer the opportunities they were looking for and so they moved back home. Nowadays – even after the welfare reforms in 1996 which restricted help for immigrants – a lot of different welfare programs, from food stamps to unemployment benefit, help those who do not succeed in finding work and even attract immigrants who would otherwise stay in their home countries. Furthermore, the level of skills of 90 per cent of the new immigrants is relatively low. And so are their earnings. As a result, state governments spend between $11-22 billion to provide welfare to immigrants, and welfare costs are steadily climbing. Today welfare use by immigrants is 43 percent higher than by native-born Americans and immigrants are still entering the United States in high numbers. In many states immigration is said to be one of the main reasons for their financial problems and many experts believe that unless American immigration policies are changed there will be no end to the problem in the near future.

457 words

a Which of the following topics are mentioned in the text?
unemployment - transport - welfare costs - California - Texas - skills - welfare use - environment

b Answer the following questions in complete sentences.

1. How can immigrants help the economy
   - as workers?
   - as consumers?
2. What are many people afraid of in a bad economic situation?
3. How has the welfare state developed since 1900?
4. How has the level of immigrants' skills changed in the last few decades?

c List the costs and the benefits of immigration according to the text.

d Find more arguments and state your point of view.

e Write a comment: Immigration helps the economy. Do you agree?

f Translate lines 44 to 70.

Follow these steps

1.	**Read the whole text first, even if you only have to translate part of it. Make sure you understand the unknown words. Use your dictionary.**	decade = Jahrzehnt, 10 Jahre
2.	**Translate the text into German in a way that you have a first draft. (word-by-word translation)**	Jedoch in den letzten Jahrzehnten haben sich die Dinge geändert.
3.	**Work through the text again. Compare the structure and the meaning of the English and the German sentence. Try to produce a suitable German final version.**	In den letzten Jahren hat sich jedoch die Situation geändert.

Info-Box: How to write a translation

Remember:

1. Read and try to understand the complete text.
2. Look up unknown words in your dictionary.
3. Make a first draft in a word-by-word translation.
4. Go through the text again and produce a final version in good German.

1 The new majority

The Californian lifestyle has grown out of many different cultures. People have been coming here from other American states and from foreign countries for a long time now. They have come to find better jobs, a better house, a better climate – in short a better way of life.

a Listen and put the photos into the correct order.

b Find out where these people came from.

c Find out where they live now in California.

d Listen again and find out the following information about the people:

	Marital Status	Language	Job in U.S./home country	Reasons for emigrating	Problems
Santiago					
Lev Pischnik					
Mom Chay Suth					

2 Minority groups

In many countries minority groups often face problems. Therefore some people are against immigration. On the other hand, immigration brings a lot of benefits to a country.

a Make a list of all the benefits you can think of in the following areas.

1. economy
2. employment
3. welfare
4. education
5. culture
6. language

b A local newspaper has just published a letter from someone wanting to stop immigration.
Write a letter to the editor of the newspaper, giving your arguments for immigration.

Technical topics

Back to the drawing board

Imagine you are an engineer at a major car maker in charge of creating a new family car. What aspects would you have to think about? Clearly cars should not pollute the environment so much, but they should also be cheap, comfortable, safe and above all fun to drive. These seemingly contradictory demands have sent the vehicle designers back to their drawing boards and computers.

Let us imagine you have been given a design envelope that includes a number of typical guidelines to follow: the car must have space for 5–7 people, good acceleration, and, of course, an economic fuel consumption. You must build in safety devices such as air bags and crush zones and side impact beams. The car must come with a navigation system and an electronic cruise control. And do not forget the vehicle must be built – and sold – at a reasonable price. Another significant guideline that you have to follow is based on tough environmental regulations.

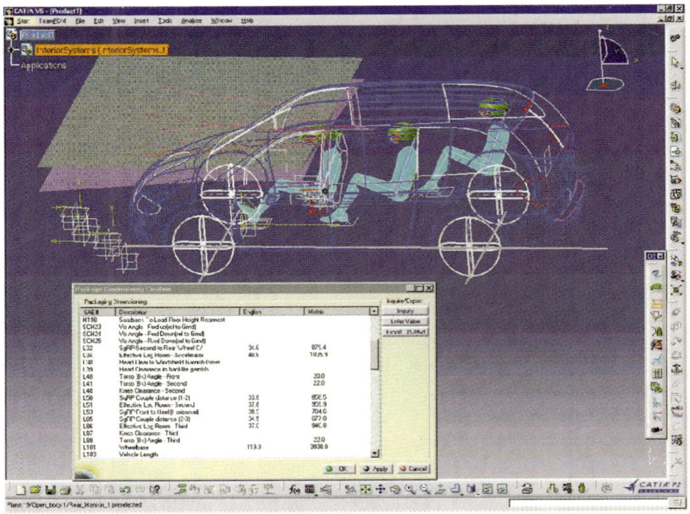

Green has become the new colour of the automobile market. As a production manager of a leading car company says, "Today 10 % of our investments are directly put into environmental protection." What must evolve is a new generation of vehicles which reduce carbon dioxide emissions and at the same time increase fuel efficiency. Around the world car makers are experimenting with a range of new vehicles. The hybrid vehicle, which has both an internal combustion engine and an electric motor, is one. However, models which are based on fuel cell technology are proving even more popular with many designers. Other companies are looking into the possibilities of producing vehicles which run on solar power or alternative fuels such as hydrogen, methane or even rape seed. The company which discovers a car which is practical, affordable and runs on a non-polluting power source could be as successful as Henry Ford was when he standardized the production of the Model T.

Some of these "future" cars are already coming off the assembly lines, but most of the car makers prefer to design first so-called "concept cars". These are forerunners of intended production vehicles. After the vehicles have been carefully designed to all the guidelines, they are shown at automobile shows around the world to test the public's reaction. These concept cars give drivers a taste of the future. Moreover, they demonstrate that industrial design need not just follow the demands of the present market, but these cars also prepare the ground for future vehicles and produce a sort of suggestion box on wheels. If the public's reaction is positive, the manufacturers then start to use the concept car to test the limits of the technology available.

Over the years many significant developments have been made, especially in the field of electronics. Anti-lock brakes, electronic stability control and digital motor electronics are just three examples. Nevertheless, the main challenge the car makers and their designers will face in the future will be in the area of emissions, fuel efficiency and safety without losing the flair and spirit of individual mobility that we have all got so used to. So back to your drawing boards you dream-machine designers! **523 words**

Exam preparation

Comprehension questions

After you have read the text carefully, answer the following questions in your own words as far as possible.

1. Which aspects do car designers have to consider nowadays?
2. Why have car companies become environmentally conscious?
3. Which new developments, according to the text, are being made in the field of emission reduction?
4. What important role do concept cars play in the automobile industry?

Summary

Using your own words, give a summary of the text. It should be about 250 words long.

Vocabulary

1. Explain the following:
   a) guidelines (line 12)
   b) internal combustion engine (line 32)
   c) fuel efficiency (line 29)
   d) a suggestion box on wheels (lines 54/55)
   e) individual mobility (line 65)
2. Look in the text to find opposites for the following words:
   a) expensive
   b) danger
   c) destruction
   d) increase
   e) unimportant

Translation

Translate lines 58 to 67.

Comment

Comment on the following statement, using your own knowledge and ideas.

"Further technical developments in the car industry will solve all our future transport problems."

Grammar revision

1 Mixed tenses

■ Put the verbs in brackets into the correct tense.

Dear Alex,
I'm writing to you while on my journey to Scotland. Yes, I'm on the train! I finally **1. (sell)** my car last year because of the damage cars do to the environment. Until then I **2. (use)** my car to visit my parents in Edinburgh, but last time I **3. (go)**, the journey **4. (take)** 15 hours and we **5. (not arrive)** until midnight. The next day I **6. (fall)** asleep while we **7. (watch)** the Scotland versus England rugby game, which was the reason I **8. (go)** there in the first place. Travelling by train is definitely quicker and more environmentally-friendly and now at least I **9. (know)** that I **10. (arrive)** fresh and cheerful.
At the moment Graham is with me and we **11. (enjoy)** every minute. Perhaps there **12. (be)** time to go out for a meal in a restaurant with my parents this evening. We **13. (not manage)** to do that for many years now. My father **14. (get)** old and he **15. (not go)** out very often. Tomorrow, we **16. (visit)** an old friend of ours in Glasgow. We are near York now. The countryside **17. (look)** beautiful round here. I **18. (not have)** such a pleasant journey since I **19. (be)** at university. That was while I **20. (study)** in Bristol. At that time my parents **21. (live)** in London and I **22. (use)** the train to get home in the holidays.

Best wishes
Paul

2 Passive

■ Rewrite the parts in bold type in the passive form.

Green cars
At an International Automobile show last autumn, **people ignored the usual talk of acceleration rates**. Instead, **car makers promoted environmentally-friendly cars**. Finally, **the car industry is reducing the amount of time and money spent on fast cars** in order to protect the environment.
As governments increase efforts to reduce air pollution, more and more car makers have produced 'green' cars. And in the future, **they will introduce more cars with zero emissions and recyclable parts.**
Electric cars which run on batteries and do not pollute the environment are particularly popular. With some of these cars, **you can recharge the batteries** in only five minutes. Nowadays too, **you don't need car scrap heaps** anymore, as **car companies take back and recycle their cars**. Big changes indeed in a big industry!

The internal combustion engine

a Look at the following drawing of an internal combustion engine and read the accompanying text.

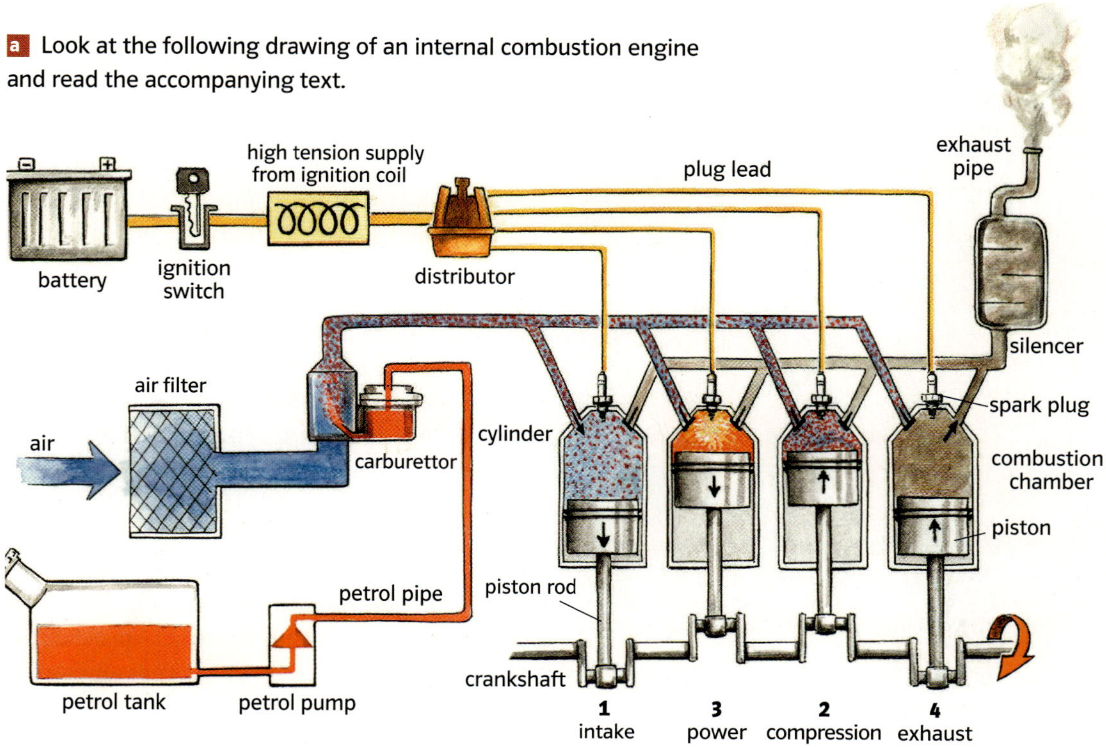

The function of the ignition system is to provide a spark in the combustion chamber to ignite the air/fuel mixture at the correct time. The system consists of an ignition switch, a battery, a coil, a distributor, spark plugs and leads.
The coil produces the high tension (high voltage) that is required to make the spark at the spark plugs. The distributor has the function of distributing the high tension supply to each spark at the correct moment. The spark plugs are screwed into the cylinder head at the top of the combustion chamber.

b Translate the above information into German.

c In the same way as above explain how the fuel system works.

d Describe the four stages in a four-stroke engine with the help of the words in the box.

> move up/move down/push out/
> compress/suck into/mix/create/
> turn/ignite/repeat

Example:
In the intake stroke the piston moves down and the petrol/air mixture is sucked into the cylinder. In the compression stroke …

e What would you suggest?
It is a cold damp day and your car won't start. The engine turns over quite well but it sounds lifeless. There is a smell of petrol, too.

Energy production

Energy from the wind

When a small wind power company decided to build a wind farm near his home in Wales, Terry Williams was one of the first to welcome the decision. However, that was several years ago. Once the wind turbines started working, his opinion changed. "It's like being near a large air-conditioning unit," he says, "the noise comes from all directions. In some rooms you can hear it even with the double-glazed windows shut. And if it gets windy in the night, the turbines start and wake us up."

Other people living in the area also noticed the noise, which is caused partly by the turbine blades and partly by the sound of the gear mechanisms. But for houses 600 metres or more from a turbine, this is not such a big problem as noise levels there are little above the natural background level of 19 decibels. But noise is not the only problem connected with wind farms. The appearance is also a problem as some people consider that they are destroying the beauty of the Welsh countryside. Because of this, a group of experts have been examining the effects of existing wind farms. What they have found is that smaller wind farms needn't be an ugly addition to the landscape. The larger installations, however, can be seen more than nine miles away. This, the experts agree, spoils the view of the countryside.

Yet many of the people who live and work near the turbines are not disturbed by them. In one survey, 70 % of the people in villages near wind farms said they would be happy to see more of them. Some even said, "Put them anywhere – they are pretty, they look like daffodils – they should be painted yellow."

At present, wind energy generates only a small amount of energy. Several hundred turbines in Wales, where 40 % of all the UK's turbines are situated, produce 300 kilowatts each when they are turning. Although their output is less than an average gas- or coal-fired power station, the potential is there. Plans for the biggest off-shore wind farm in the UK have already been approved. This wind farm off the north Wales coast will provide electricity for 50,000 homes. According to a report by the Welsh Assembly in 2002 the country needs many more wind farms and should aim to produce enough electricity from renewable sources to supply more than three million homes by 2010.

Countries, such as Denmark have realised that wind energy saves resources and produces neither dangerous emissions nor the toxic waste that nuclear power plants produce. So they have gone ahead with the development of wind farms and are already successfully producing larger quantities of energy from wind power.

Although the British government has not always been willing to support the growth of renewable forms of energy in general, its goal is now to produce more energy from renewables, wind, wave, geo-thermal or hydro-electric power in the next few years – 20 % by the year 2020. However, only a small proportion of its electricity tax money is spent on renewables, most of it goes on nuclear power. There could be a quick and easy solution to the problems which wind energy, like every new technology, faces at the beginning. If the government spends more money on wind turbines, their effect on the landscape – both from the visual and the noise point of view will quickly be reduced and Terry Williams will enjoy the benefits of environmentally-friendly energy and sleep well again at night.

520 words

Exam preparation

Comprehension questions

■ After you have read the text carefully, answer the following questions in your own words as far as possible.

1. Why has Terry Williams' attitude to environmentally-friendly technology changed?
2. What advantages do wind farms have compared with conventional power stations?
3. What, according to the text, are the negative aspects of wind farms?
4. What action is needed to eliminate the early problems in wind power technology?
5. What plans are there for the development of renewable energy sources in the UK in future?

Summary

■ Using your own words, give a summary of the text. It should be about 250 words long.

Vocabulary

1. Explain the following:
   a) wind farm (line 2)
   b) natural background level (line 16)
   c) double-glazed windows (line 8)
   d) turbine (line 5)
   e) hydro-electric power (line 58)
2. Look in the text to find the opposites of these words:
   a) beautiful
   b) small
   c) unwilling
   d) increase
   e) far away
   f) end
   g) badly
   h) disagree

Translation

■ Translate lines 33 to 46.

Comment

■ Comment on the following statement, using your own knowledge and ideas.

"Alternative energy? – No, thank you."

Grammar revision

1 Mixed tenses

Put the verbs in brackets into the correct form.

Dear Patrick,

Recently I **1. (join)** the organization called 'Friends of the Earth'. I **2. (not know)** if I **3. (tell)** you this while I **4. (stay)** with you a couple of weeks ago. It's a group which **5. (aim)** to protect the environment.

I **6. (expect)** you **7. (hear)** of them already. Anyway, I have decided that I **8. (attend)** one of their meetings next week. Would you like to come with me? The subject which they **9. (discuss)** at this meeting is the energy supply in the Third World. It **10. (seem)** that they **11. (find)** it very expensive to supply remote villages with energy from a grid supply and so in future they probably **12. (use)** renewable energy sources.

I **13. (send)** you a copy of their leaflet if you like, so that you can read about it yourself. Until I **14. (read)** it last week, I hadn't known much about the problem. I certainly **15. (not realise)** before that how many people in the Third World live without an energy supply.

Apparently 70% of the population of the developing world **16. (not have)** any electricity at all. But in the last few years some Third World governments **17. (start)** to build solar, water and wind power installations. The main problem is that these installations also **18. (cost)** a lot of money. To help pay these costs, at present some renewable energy companies in the industrialised world **19. (organize)** special funds. The World Bank believes that developing countries **20. (invest)** $1 trillion in power systems of one sort or another in the next ten years. You can learn more about it next week.

See you at the meeting?

Sally

2 Conditionals

Put the verbs in brackets into the correct tense.

Two engineers are talking about wind energy.

If the government **1. (not support)** nuclear energy in the way they do, there **2. (be)** much more money left to develop environmentally-friendly methods of energy production, like wind energy.

I know, if the government **3. (continue)** to spend so little money on renewables, the technology **4. (not develop)** fast enough and that means critics **5. (have)** even more time to find arguments against it.

Yes, it makes me furious. If they **6. (start)** to build wind farms years ago, by now the problems **7. (solve)**. A colleague of mine who lives in Wales says that if they **8. (have)** a coal-fired power station near them instead of a wind farm, he **9. (find)** it much uglier. The wind farms look pretty, he thinks.

Electric supply

Here is a diagram of an energy supply system from power station to consumer.

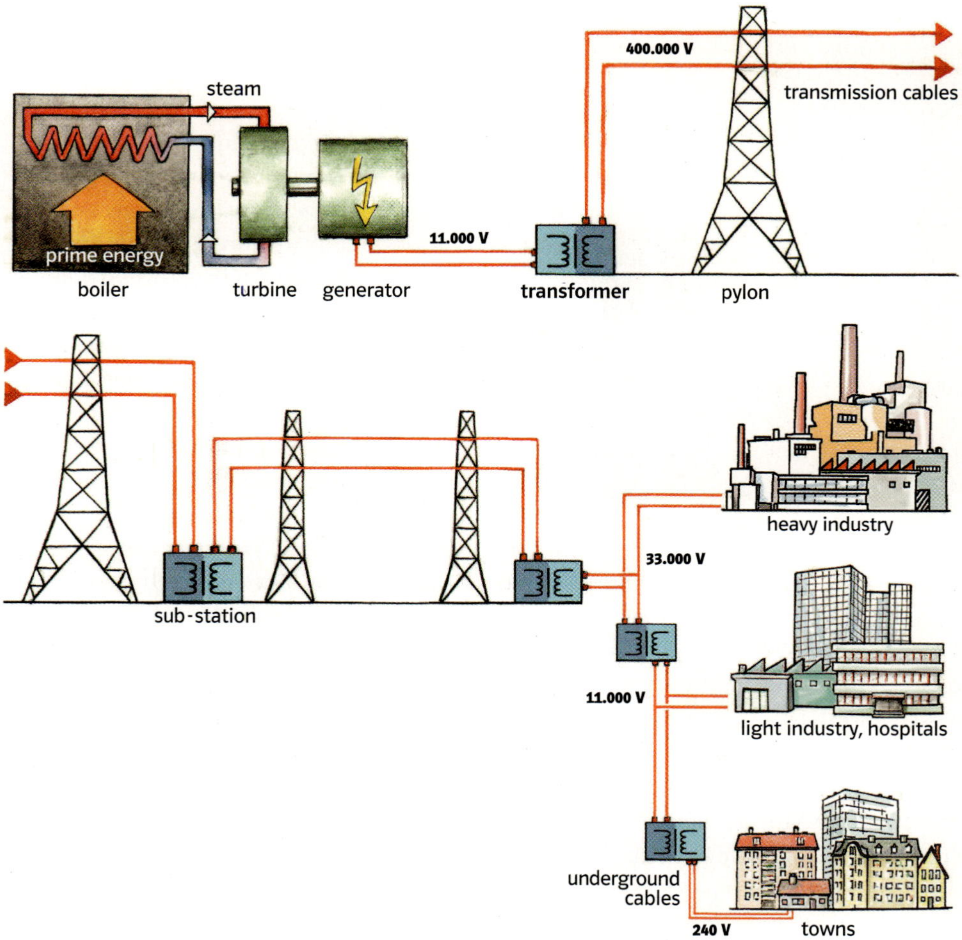

Electricity Boards are part of the electricity supply industry of England and Wales. The industry supplies power to nearly 50 million people. It is responsible for developing and running an efficient and economical system of electricity supply. It is one of Britain's biggest businesses and its power system is one of the largest in the world.

a Describe how electricity is produced.

b Describe how electricity is supplied to the consumer (households/factories).

c When electricity is supplied, the voltage is changed at various points.
1. Give examples of such changes.
2. Why are these changes necessary?

d Translate the text.

Technology and pollution

A future for the earth?

"One generation passes away, and another generation comes: but the earth lives on forever." (Holy Bible)

No, not forever. For man has now reached a point in his evolution where he has the power to influence the present and the future state of the planet, for better or for worse.
Since the beginning of the Industrial Revolution, smokestacks have released harmful gases into the atmosphere, factories have dumped toxic wastes into rivers and streams, cars have consumed irreplaceable fossil fuels and polluted the air with their exhaust gases. In the name of progress, forests have been cut down and lakes and oceans poisoned with pesticides and chemicals.
The continuing destruction of the atmosphere's ozone layer is partly caused by the use of atmosphere-destroying chlorofluorocarbons (CFCs) in different types of foam, insulation and air-conditioners. Perhaps most threatening of all, the destruction of the tropical forests, home to at least half the earth's plant and animal species, continues at a rate equal to one football field a second. The resulting worldwide extinction of species promises to be at least as great as the mass extinction that happened at the end of the age of dinosaurs.
For a long time scientists have pointed to the damage and its possible consequences. But no one has paid much attention. Now things have become so bad that people are beginning to listen. Surely everyone must now notice that the Earth is in danger. In many countries heat waves every summer now drive temperatures above 40 degrees Centigrade, raising fears that the "greenhouse effect" – global warming – is already under way. Because of the lack of rain, some forests, including the Yellowstone National Park in the U.S., have already gone up in flames several times. And on many beaches, garbage, raw sewage and medical waste spoil the fun of bathers and confront them personally with the growing pollution of the oceans. Pollution in Europe has closed beaches on the Mediterranean, the North Sea and the English Channel. Thousands of seals have died because their immune systems have been weakened by pollution.
What will happen, if nothing is done? The increase of CO_2 in the atmosphere could raise the planet's average temperature by 2 to 6 degrees Centigrade within the next fifty years. That could cause the oceans to rise by several metres, flooding coastal areas and ruining huge parts of farmland. Changing weather conditions can make large areas infertile and uninhabitable, resulting in migration on a scale which we have not experienced up till now.
Let there be no illusion. Although several world conferences (Earth Summit: Kyoto 1997, Buenos Aires 1998, Bonn 2001, Johannesburg 2002), have taken place to discuss global warming, and despite the fact that many companies have already been able to cut their emissions of carbon dioxide considerably, a lot more must be done. Taking effective action to halt the massive damage to the earth's environment will require a mobilization of political will, international cooperation and sacrifice unknown except in wartime. But humanity is at war right now, and it is no exaggeration to call it a war for survival. It is a war in which all nations must be allies. Both the causes and effects of the problems that threaten the earth are global, and they must be attacked globally.
Every individual on the planet must be made aware of the serious situation and of the need to do something. No attempt to protect the environment will be successful in the long run unless ordinary people – the American housewife, the Chinese factory worker, the African farmer and the European office worker – are willing to change their lifestyles.

586 words

Exam preparation

Comprehension questions

After you have read the text carefully, answer the following questions in your own words as far as possible.

1. How has man, in the name of progress, changed the earth's environment?
2. How has nature reacted to such treatment?
3. What will happen if nothing is done about this serious situation?
4. Why does the author say that humanity is at war right now?

Summary

Using your own words, give a summary of the text. It should be about 250 words long.

Vocabulary

1. Find opposites for the following words taken from the text:
   a) forever (line 1)
   b) better (line 3)
   c) much (line 26)
   d) harmful (line 6)
   e) irreplaceable (line 8)
   f) destruction (line 13)
   g) danger (line 28)
   h) weaken (line 41)
2. Look in the text to find the corresponding verbs for these nouns:
   a) life
   b) poison
   c) threat
   d) confrontation
   e) requirement
   f) protection

Translation

Translate lines 51 to 65.

Comment

Write a letter giving your ideas on the measures that governments, industry and ordinary people can take in order to improve the environment. Imagine the letter is addressed to the editor of the magazine which published this article.

Grammar revision

1 Mixed tenses

■ Put the verbs in brackets into the correct tense.

Feeling the Heat

For more than twenty years now, scientists **1. (warn)** about the gases from cars and factories. This **2. (produce)** massive climate changes in the next few years. The temperature of the earth **3. (already/rise)** by several degrees. Years ago nobody **4. (pay)** much attention. But then NASA's Institute for Space Studies **5. (turn)** global warming into front-page news. James Hansen, head of the institute, **6. (say)** that global warming **7. (change)** from theory into fact. Since then the amount of CO_2 in the atmosphere **8. (increase)** everywhere. So if we **9. (not/change)** our energy consumption habits, our planet **10. (become)** uninhabitable. For that reason many countries **11. (already/introduce)** higher petrol taxes, and others **12. (plan)** a CO_2 tax on all fossil-fuel energy sources.

2 Infinitive or "ing"-form?

■ Fill in the infinitive with or without "to" or a suitable "ing"-form.

A hundred-year plan for the future

Japan has made impressive steps on its way to become a First World leader in **1. (control)** air pollution, in **2. (reduce)** energy consumption and in **3. (provide)** its citizens with clean water. By **4. (pass)** the Basic Environmental Law **5. (include)** a national environmental plan and **6. (establish)** a Central Environmental Council, Japan has opened a new round of "green" laws. This will not only make companies **7. (reduce)** emissions, but it will also help them **8. (develop)** clean energy sources.
Most Japanese companies are now interested in **9. (become)** environmentally correct. For example Chiyoda Corp. has developed a method of **10. (clean)** power plant exhaust. After **11. (work)** on it for several years Nippon Steel Corp. has announced a new waste-melting system. This process will reduce the volume of toxic waste by **12. (convert)** most of it into recyclable material.
On the other hand critics still point out that, for example, the country's industry has caused a lot of damage by **13. (use)** wood **14. (come)** from the tropical forests in Southeast Asia. When we asked him **15. (give)** a statement, Saburo Kato, president of an environmental research institute, said: "There are still some difficulties in **16. (cooperate)** on an international level, but we aren't afraid of **17. (face)** competition in the environmental field."

Waste management

Millions of electric and electronic appliances are produced every year. But they have a short life. TV-sets are often replaced after seven or eight years, personal computers are technically out-of-date soon after they have been bought. So what to do with the rubbish?
Today more and more firms are trying to avoid or substitute harmful materials and to reduce the amount of scrap.

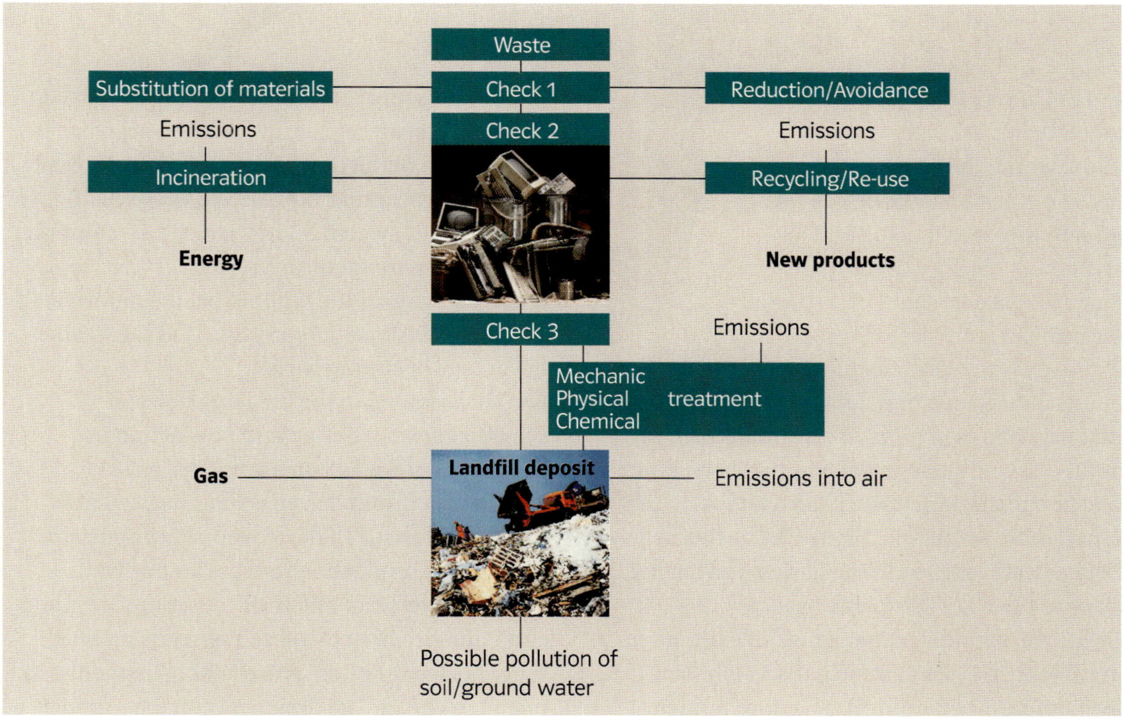

a Describe the decisions you should make at the different checks in the diagram above. Use the words in the box.

clean, recharge, maintain, burn, avoid, substitute, reduce, treat, repair, melt, re-use, recycle, dump

Example:
At check 1 you must decide if you can substitute materials or …
At check 2 you …

b What would you do with the following parts of electronic appliances? Think of reducing, substituting, recycling and incinerating.

1. plastic parts
2. paint
3. cardboard packaging
4. oil
5. batteries
6. mixture of plastic, metal, glass

c Give reasons for and possible disadvantages to your suggestions.

Unit

14 CAD/CAM/CIM

Automating design

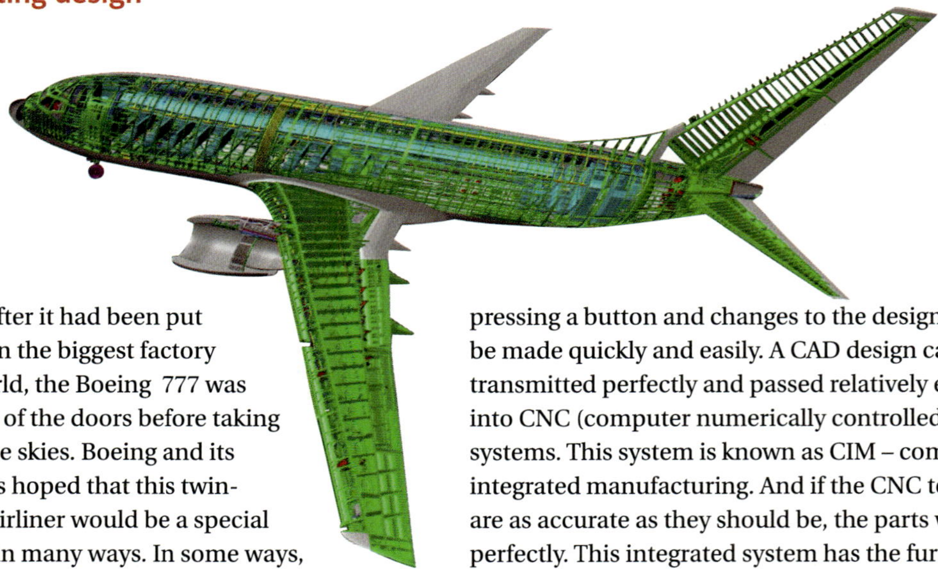

In 1994, after it had been put together in the biggest factory in the world, the Boeing 777 was rolled out of the doors before taking off into the skies. Boeing and its customers hoped that this twin-engined airliner would be a special machine in many ways. In some ways, it already was. Normally the company builds models of a new aeroplane to make sure that its millions of parts fit together. However, in this case no such prototypes had been built. The first plane which was produced was a fully functioning aeroplane. Using CAD technology Boeing were able to cut the design and testing stage of their 777 aeroplane by 90%.

Of course, a model of the plane did exist, namely in trillions of bytes of information stored in the company's computers. The development of the new model had been the biggest single test of computer-aided design (CAD) at that time. By using the computer's designs of the parts, engineers with the help of their computer programs checked that the whole thing would fit together before any pieces were made. Their checking proved to be extremely successful, as all the parts fitted to within a thousandth of an inch.

CAD programs are not spectacular, but they give the designer new tools and allow companies to change the ways they organise design. Good design is the key to sucessful manufacturing.

The advantages of automation have always been time and accuracy. CAD has both of these. Technical drawing of designs takes a long time and research shows that CAD programs can cut the time it takes to draw a design by as much as 90%. Perfect copies can be made by simply pressing a button and changes to the design can be made quickly and easily. A CAD design can be transmitted perfectly and passed relatively easily into CNC (computer numerically controlled) tool systems. This system is known as CIM – computer integrated manufacturing. And if the CNC tools are as accurate as they should be, the parts will fit perfectly. This integrated system has the further advantage of cutting costs.

CAD has also led to the breaking down of the wall between design and production in a process known as 'concurrent design'. The idea of concurrent design is that the designers work closely with all those who play important parts in the production process. In this way, manufacturers can follow the design process and can influence it. In the past, there was no need for concurrent design as everything happened in one man's head. The designer built the product or at least oversaw its building. The aim of a good concurrent design team is to function together as one man's mind and to consider all the sides of a process at the same time. However, to get several heads within one company working together as one is not always easy. They all have to communicate with each other very clearly. They also have to know when a change has been made and what its effects might be. This is what integrated CAD systems can provide. They allow designers to see the effects of changes in ways that were not possible in the past. So, as well as a design tool, CAD is also a communications technology which ties all the knowledge in the team together, allowing them to plan ahead and make the right decisions before it's too late.

571 words

Exam preparation

Comprehension questions

■ After you have read the text carefully, answer the following questions in your own words as far as possible.

1. What difference is there between the Boeing 777 and any aeroplane built before that?
2. What influence does CAD have on the manufacturing process?
3. Describe briefly how the CIM system functions?
4. What improvements does the use of integrated technology have both for the product and the company?

Summary

■ Using your own words, give a summary of the text. It should be about 250 words long.

Vocabulary

1. Explain the following:
   a) twin-engined airliner (lines 6/7)
   b) prototype (line 12)
   c) CNC tool systems (lines 41/42)
   d) manufacturers (line 53)
   e) computer integrated manufacturing (CIM) (lines 42/43)

2. Look in the text to find synonyms for these words:
   a) largest
   b) to assemble
   c) to make sure
   d) to shorten
   e) fast
   f) goal
   g) brain
   h) to talk to each other

Translation

■ Translate lines 32 to 46.

Comment

■ Comment on the following statement, using your own knowledge and ideas.

"Computers are steadily improving the quality of our work and working conditions."

Grammar revision

1 Mixed tenses

Put the verbs in brackets into the correct form.

Dear Sam,
I'm sorry that I **1. (not write)** to you since your last email. But last month my parents **2. (buy)** me a new computer and since then I **3. (be)** very busy with it, finding out the new possibilities. As I now **4. (have)** many more possibilities with the new system, I **5. (try)** to link up with other computer users all over the world. I am sure, as you **6. (get)** your new computer a few months ago, that you already **7. (contact)** a lot of other computer freaks. If you **8. (like)**, we **9. (can exchange)** some addresses and some interesting homepages that we **10. (find)** up till now.
By the way, before I **11. (get)** the new computer, my father and I **12. (not get on)** too well together, but now we have a common interest and lots of new friends. We **13. (have)** a lot of fun together at the moment. Sometimes we **14. (take)** part in very interesting chats on-line. Last week we **15. (talk)** to a man who **16. (work)** on a pipeline in Alaska. He **17. (not get out)** much there in those cold conditions. Just imagine, soon we **18. (have)** friends all over the world.

Best wishes
David

2 Modal auxiliaries

Put the correct modal or its auxiliary form into these sentences.

Visual engineering
Car makers have recently introduced a new system of manufacturing engineering, called visual engineering. It means that designers **(1)** draw their designs by hand anymore. In future, sophisticated computers **(2)** build solid models on the screen for them instead. Up to now, designers **(3)** use old-fashioned design and development techniques with drawings and real models. But now already they **(4)** assemble parts on the screen to see if they work. In this way car makers **(5)** cut the time needed to develop the product considerably. In order to keep up with these developments in the manufacturing process, firms **(6)** make other changes too. The office work **(7)** slow down the process or the manufacturing advantages will be lost. There are still some problems to be solved, but soon car makers **(8)** design cars electronically from start to finish.

1 Hardware

a Match the names of the devices in the box with the pictures.

DVD drive • scanner • headset • keyboard • monitor • printer • mouse

b Describe the function of these hardware parts. Use the words in the box.

input/output device • cursor • screen • store • print • display • show • move

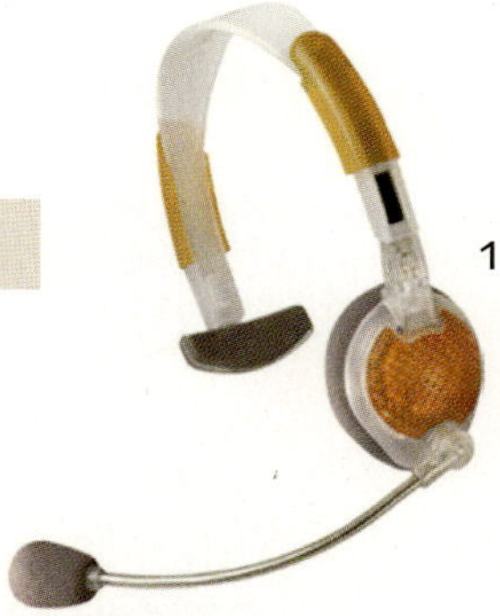
1

2

3

4

5

6

7

2 CAD

■ Translate the following text.

Implementing a CAD program is an enormously complicated task for a computer, which is why CAD use was limited until recently to large computers.

Work with a large computer is as a rule, organized in such a way that one central computer supplies several terminals with the required data. As a result of the large quantities of data which are often produced by CAD programs, the time it takes to produce an answer is relatively long and can have a negative effect on a continuous work process. On the basis of the current level of technological progress, an average answer time which is one second or more is not acceptable.

Building bridges

Building a suspension bridge is one of the biggest challenges which bridge builders face. To build a suspension bridge you need two or four big steel cables between two support towers. Between these towers there is a span of unsupported distance, which in some cases is over one kilometer long.
Moreover, you require many smaller cables, hanging down from the main cables, to hold the road or railway. Suspension bridges are the most technically difficult bridges to build, but in the case of the Golden Gate Bridge in San Francisco it was the only way. Two factors made any other kind of bridge impossible. The first was that some of the largest ships in the world used the Bay of San Francisco to enter the main harbour of the city and they needed to use the deepest part of the channel for this purpose. The second factor which influenced the engineers' decision to design a long suspension bridge was that the channel is not only deep but is also affected by strong tides and big waves in bad weather. Therefore supports in the middle of the channel could not be erected.
Work on the bridge began in 1933 after an engineer named Joseph Strauss had finished his plans for the suspension bridge. Obviously the work during construction was going to be very dangerous, so Strauss designed a big net to go under the bridge to catch workers who fell. It was also the first time that construction workers had to wear protection helmets and safety belts with ropes on a building site. The first part of the work was to build the two support towers. Both of these towers had to be built on the ground which was under the water level. Consequently, a cofferdam was constructed. When the water had been pumped out of the cofferdam, both men and machines went into the dry area and dug down ten metres below the sea level. While they were digging, the sharp edges on the bottom of the heavy cofferdam pressed down further into the ground. Then concrete was pumped in to make the foundations for the tall towers. Strauss designed the tower foundations to move under the towers in case of an earthquake. So he did not connect the towers to the concrete. In that way, the bridge was safe. Next it was time for teams of workers to put the steel beams of the towers into place. Then the towers were raised up to 227 metres above the water.
After both towers had been completed, the main cables had to be put into position. These cables needed to be both strong and flexible because they had to bend over the tops of the towers and down to the ground behind them on the land. Here they were connected to heavy concrete anchorages in the ground. Having spanned these main suspension cables across the water, smaller cables, so-called suspenders were attached to them. The suspenders hung down in tension with the main cables. Then the road support and the road itself were connected to the suspenders. The road was wide enough for six lanes of traffic. Finally workers painted the bridge to protect it from the salt water and the misty air. In 1937 the bridge was opened.

514 words

Exam preparation

Comprehension questions

After you have read the text carefully, answer the following questions in your own words as far as possible.

1. Why did engineers decide to build a suspension bridge over the Bay of San Francisco?
2. Which changes in workers' safety came with the building of the Golden Gate Bridge?
3. Why was a cofferdam used?
4. Why were the towers not connected to the concrete below them?
5. What function do suspenders have?

Summary

Using your own words, give a summary of the text. It should be about 250 words long.

Vocabulary

1. Explain the following:
   a) unsupported distance (lines 5/6)
   b) strong tides (line 21)
   c) protection helmets (line 31)
   d) foundation (line 43)
   e) in tension (line 60)
2. Find nouns for the following verbs taken from the text:
   a) require (line 8)
   b) construct (line 36)
   c) build (line 2)
   d) enter (line 16)
   e) move (line 44)

Translation

Translate lines 51 to 66.

Comment

Comment on the following statement, using your own knowledge and ideas.

"The construction of new roads, houses and bridges often destroys both the landscape and the quality of peoples' lives. Governments should stop new planning and concentrate on conserving the environment."

Grammar revision

1 Mixed forms

■ Put the verbs in brackets into the correct form.

1. (think) you ever about the work which **2. (go on)** before a house **3. (build)**? Putting up new houses on new sites **4. (require)** much careful preparation. For example it **5. (be)** always very important to find out the kind of soil you **6. (plan)** to build on. If you **7. (not do)** soil tests, building on the wrong type of soil **8. (cause)** the houses **9. (sink)** slowly. Besides **10. (test)** the soil, it is necessary **11. (measure)** the level of the ground water. Let us imagine this for one moment. If a house **12. (construct)**, **13. (use)** wooden supports which **14. (remain)** partly below and partly above the ground water level, these supports or piles **15. (rot away)** in a short time. The result **16. (be)** that the house **17. (collapse)** soon. Once the site **18. (test)** thoroughly, roads can **19. (build)** and site toilets for the builders can **20. (put up)** on or near the site. Now work may **21. (start)** with the plotting of the area on which the houses **22. (build)**.

2 Reported speech

■ Write a report on this speech by John Charlton to some visitors to Newcastle upon Tyne. Use as many different reporting verbs as you can.

The new shopping centre

"I would first like to welcome you to Newcastle. My name is John Charlton and I am Newcastle upon Tyne's planning officer. This evening I am going to give a short talk on the development of Newcastle's inner city. After lots of careful planning and public discussion we now have completed our new shopping centre right here in the city centre. We are very proud of it. It has been developed as a system of pedestrian malls. Most of these malls are at a higher level than the streets. Our city buses and service vehicles operate below the pedestrian shopping zones. Of course we didn't forget the cars which many people use to travel to the city centre. Multi-storey car parks have been built at the edge of the central area. We, as planners, had two main objectives while we were designing this project. On the one hand, we didn't want to destroy the existing historical buildings and on the other hand we wanted to concentrate the shopping facilities in a more convenient and compact area. As you can see from this drawing shoppers can do all their shopping in the dry and are not bothered by the noise and fumes of the traffic. If you have any questions, don't hesitate to ask. Yes. The lady over there in the red dress. Madam, what's your question, please?" …

A new highway

When civil engineers planned the U.S. Interstate Highway System, they had to make some difficult decisions. In planning a highway between Adamsville and River City, engineers thought of three plans.

- Look at the drawing of the three plans and consider the advantages and the disadvantages of each plan. The notes below will help you.

- Then write a report describing the three plans and decide which of the plans you prefer. Give reasons.

Plan A:
- lower hills and raise valleys
- build bridge over South Road
- build bridge over the West Branch River
- destroy homes in the Lower River City
- use 16 kilometres of pavement
- spend $15 million

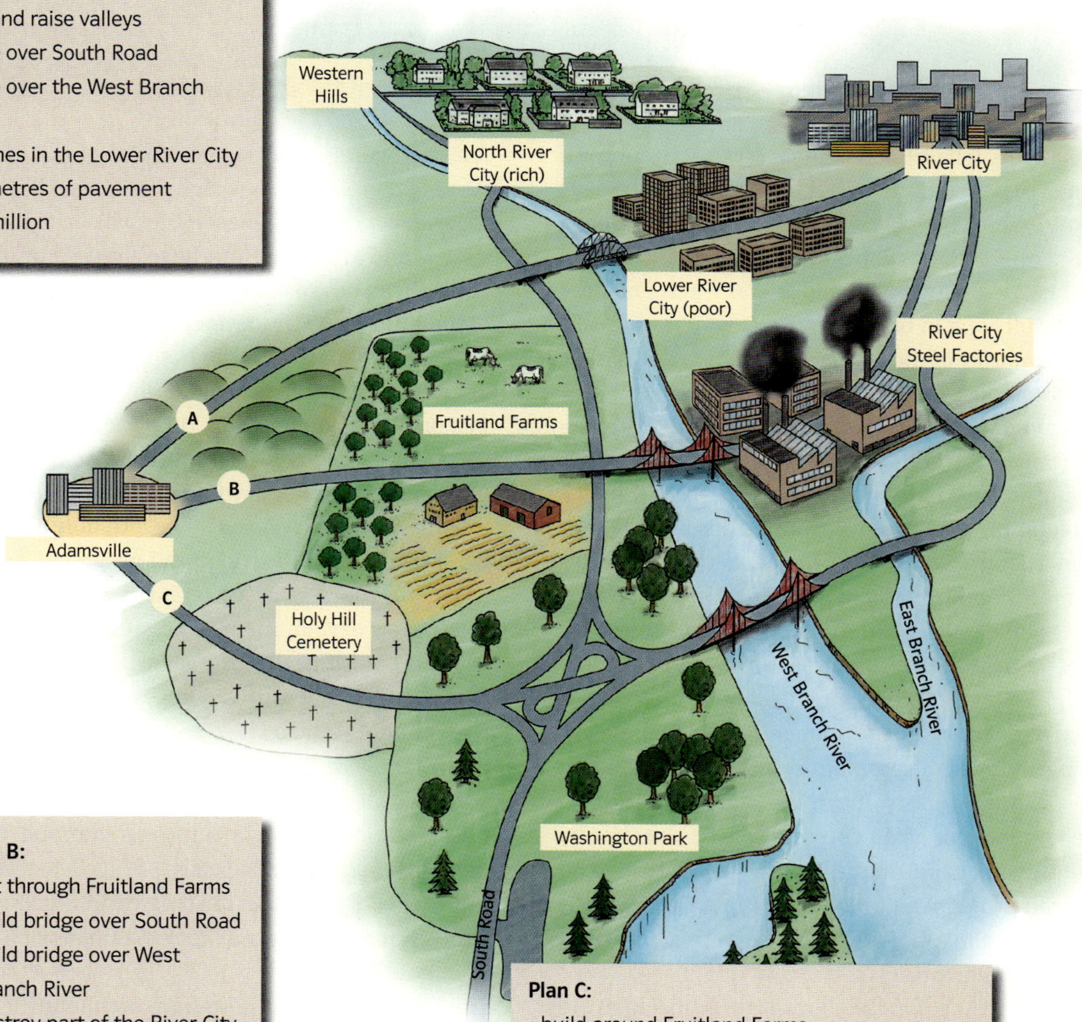

Plan B:
- cut through Fruitland Farms
- build bridge over South Road
- build bridge over West Branch River
- destroy part of the River City Steel Factories
- use 24 kilometres of pavement
- spend $25 million

Plan C:
- build around Fruitland Farms
- cut through Holy Hill Cemetery
- cut through Washington Park
- build a flyover over South Road
- build bridge across the West Branch River
- build two bridges across the East Branch River
- build around the River City Steel Factories
- use 35 kilometres of pavement
- spend $34 million

Unit

16 Interconnected technology

The home of the future

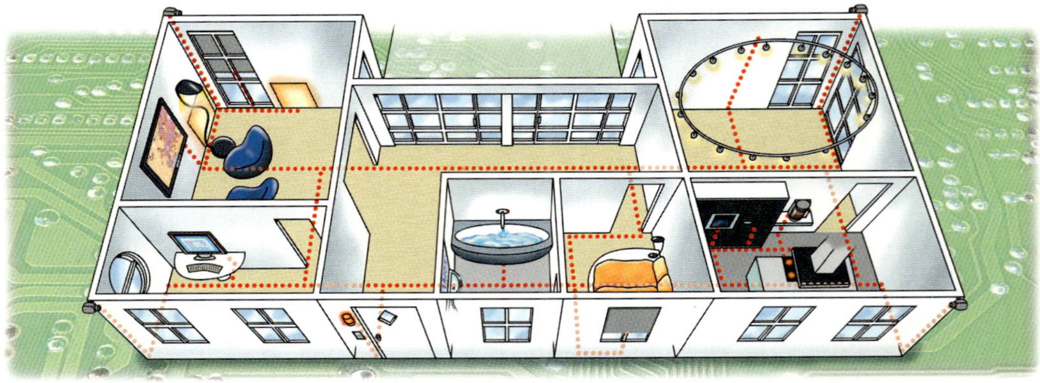

If changing television channels with a remote control seems to be too much effort, say hello to the smart home where you can run a bath, make the coffee, check on the baby, unlock the door and of course change TV channels using just your voice, and all while still lying in bed.

For the family of the future, the smart home has all this interconnected technology and more. You can also program the temperature of the bath water and operate the curtains and lighting remotely. Every room can have its own radio or TV channel with different programmes transmitted at the same time from a central computer. There is also the chance for members of the family to play PC games against each other while in different rooms. And while brushing your teeth in the morning the mirror over the sink can give you the latest news headlines. You don't even need to be in the house to operate the technology. Using an internet compatible phone while you are on your way home, you can also start a series of preparations for when you get there. In this way, by just pressing a button the heating can be turned on or you can program a number of things, e.g. the kettle to boil water, the robot dog to bark and the hifi to play some suitable music.

And how is it all possible? From a time in which only your PC was connected to the telephone line, we are now moving towards a time in which all the devices in the home can talk to each other and to us. The key to it all is a wireless connection system such as Bluetooth – named by the way after a tenth century Danish king with a bad tooth! Systems such as this use short-range radio signals on the global frequency band to connect various digital devices. They allow the transfer of data between these devices at high speeds and at present they have a range sufficient to cover the needs of most houses. So in the smart home of the future all IP (internet protocol) compatible products will use RF (radio frequency) technologies to link up to the home server.

Just one question remains. Do we really need all this interconnectivity? We already have the first interconnected fridge whose cool box is filled with intelligent features. Its users are not only able to check what foods are in the fridge and program it to order more, but can also access recipes which use those foods. New developments suggest that you could also use the fridge to surf the net, download music files, email digital camera images and have speakers, a microphone and a video camera so that members of the family can leave each other messages. Plus it could also have a radio and TV so you don't have to miss a second of your favourite programme. There's no doubt that technologically a fridge like this is very impressive, but back in the real world will anyone really want to check out the latest financial news via their cool box?

The last couple of decades have seen many technologically impressive ideas that failed because they had too many buttons and were too complicated. Will the engineers of the future be able to develop technology which will transform our lives in a really user-friendly and helpful way?

547 words

Exam preparation

Comprehension questions

After you have read the text carefully, answer the following questions in your own wordsas far as possible.

1. What features of the smart home are different to today's homes?
2. Explain what is meant in the text by the expression 'interconnected technology'.
3. How does the smart home technology work?
4. What doubts do the authors have about smart house features?

Summary

Using your own words, give a summary of the text. It should be about 250 words long.

Vocabulary

1. Explain the following:
   a) remote control (line 1)
   b) internet compatible phone (line 19)
   c) wireless connection system (line 30)
   d) cool box (line 43)
   e) user-friendly (line 63)
2. Look in the text to find verbs for the following nouns:
   a) connection
   b) program
   c) operation
   d) suggestion
   e) transformation
   f) failure
   g) development

Translation

Translate lines 7 to 25.

Comment

Comment on the question whether interconnectivity will transform our lives in a really user-friendly and helpful way.

Unit

17 Irrigation and water shortage

A way out of the world water crisis?

A worldwide water shortage is likely to get worse over the next 25 years with billions of people affected by a global crisis. Already about 450 million people in 29 countries face serious water shortages which affect the wealthiest and the poorest nations. Expanding economies and populations are competing for fewer sources of fresh water.

According to Frank Richman, director of the International Water Management Institute, the demand for water will leave some of the world's major rivers dry throughout most of the year. "Urban centres will experience water shortages. But the rural poor will suffer the most serious consequences. Many people already do not have access to drinking water and to the quantity and quality of water that is needed to grow food."

Agricultural scientists say that farmland has to expand by up to 20 per cent over the next 25 years to feed a rising population. But environmental scientists have warned that water consumption has to fall by at least 10 per cent to protect rivers and lakes.The crisis is made worse because the freshwater is often polluted by industrial waste and by increasing amounts of raw sewage. This is the main cause of diseases in some countries. The Water Institute predicts that by 2025 a third of the world population will live in regions that suffer from regular and severe water scarcity.

Large areas of the world already face shortages today, including the Mediterranean region, Pakistan, parts of India and China, most of southern Africa as well as parts of North and South America.

A report published earlier this year found that freshwater sources were decreasing very quickly. The report says: "The amount of water taken from rivers and underground reserves has grown by 2.5 to 3 per cent annually since 1940. Water tables are falling on every continent. Moreover, the report says that freshwater sources are becoming contaminated very quickly. One reason for this is the build-up of salts in the soil caused by irrigation. Very often this leads to a point where crops can no longer be grown. So many countries that rely on irrigation to grow crops are now importing food."

William Cosgrove of the World Water Council says: "It is clear that the use of irrigation must be limited. In developing countries, irrigation today accounts for more than 80 per cent of the water consumed. Half the irrigated land of Asia is used to grow rice. 2,000 tons of water are needed to grow just one ton of grain. By 2025 more than half of the world's population will depend on rice as its main food. In parts of Africa, mangrove swamps are being cut down to make way for rice production. But after five years the soil becomes infertile and new mangroves have to be cut down.

Falling groundwater levels are not only a problem in developing countries. As the report found out ground water in many areas of the world is extracted three times faster than it is replaced. The question remains: What can be done to stop this worrying development?

518 words

Exam preparation

Comprehension questions

After you have read the text carefully, answer the following questions in your own words as far as possible.

1. What does 'world water crisis' mean?
2. What are the reasons for this water crisis?
3. What is freshwater used for?
4. Why is water shortage a particularly serious problem in developing countries?
5. How does irrigation contribute to the water crisis?

Summary

Using your own words, give a summary of the text. It should be about 250 words long.

Vocabulary

1. Look in the text to find the opposites of these words:

a) improve
b) better
c) more
d) rural
e) increasing
f) rising
g) industrialized (countries)
h) fertile

2. Look in the text to find synonyms:

a) 1000 million
b) richest
c) amount
d) illness
e) falling
f) per year
g) polluted
h) wet land

3. Find the corresponding nouns for these verbs taken from the text:

a) expand (line 19)
b) pollute (line 24)
c) publish (line 35)
d) grow (line 39)
e) develop (line 51)
f) consume (line 53)
g) irrigate (line 53)
h) extract (line 65)

Translation

Translate lines 35 to 48.

Comment/Letter

Write a letter to the editor of the newspaper and comment on the article.
Make suggestions on how this worrying development can be stopped and how we can all help.

Architecture

Reaching for the sky

Minoru Yamasaki was the architect of the World Trade Center (WTC) in New York. A second-generation Japanese American born in Seattle in 1912, he lived out the American dream in the biggest possible way – in hundreds of thousands of tons of concrete, steel, brick and glass. From a poor background he worked his way through college in Washington earning 17 cents an hour. Then he went to New York in the 1930s with 40 dollars in his pocket and got a job with the architects of the Empire State building. This was the tallest building in the world until the opening of Yamasaki's 1,360ft. twin towers of the WTC in 1973.

Yamasaki, who died in 1986, would have been horrified and surprised by the destruction of the WTC on September 11, 2001 when two planes flown by terrorists crashed into the twin towers of the WTC. Other architects however didn't find it so incredible. The engineers who built the towers had even considered the possibility of accidents caused by planes. What they hadn't really considered was the terrible heat caused by thousands of gallons of burning fuel which turned the steel supports of the building almost to butter. The upper floors collapsed and pressed with growing weight on the floors below until the entire structure imploded. However, in spite of this terrible attack architects maintain that this is not the end of the skyscraper.

The skyscraper as we know it first appeared in Chicago after the city was nearly completely destroyed by fire in 1873. This gave architects the opportunity to design new buildings using all the latest construction techniques and materials. A combination of reliable steel-frame construction and elevators, which were invented in the 1850s, allowed architects to build high above the ground. The first real skyscraper was built in Chicago's business area and had fifteen floors. It quickly became popular with architects, politicians and the general public in other cities and Manhattan especially took up the skyscraper in the 1890s – determined to do better than Chicago. From then on the Manhattan skyscraper – symbol of American values and of New York in particular, was copied throughout the world.

According to Norman Foster, the well-known architect, this trend isn't going to change in the future either. He maintains, "Skyscrapers are as much a reality as urbanisation itself. They're not going to go away. Yes, we can design to counteract forces of nature – earthquakes for example, but there's no way architects can design for such extremes as terrorist attacks. However, we don't stop sailing when a ship sinks or give up underground systems when a terrorist plants a bomb. Like everyone else, architects and engineers learn from feedback and by experience."

So skyscrapers – the name comes from the highest sail of a ship – are here to stay unless we want our cities to spread across miles of land or to be built in bunkers underground. But can we make them safer? "What we might learn as architects," says another architect, "is to take a new look at issues of fire and safety. Could we include safe areas in tall buildings, zones that are as far as possible fire-proof and crush-proof? Maybe." He believes also that getting rid of commercial centers altogether is not the answer. "People like to gather and New York is an exciting place with so many people from different cultures. You could have all the electronic communications technology you could ever want, but you would miss the city."

577 words

Exam preparation

Comprehension questions

After you have read the text carefully, answer the following questions in your own words as far as possible.

1. Why was Minoru Yamasaki's story typical of the American dream?
2. What was the view of engineers about possible accidents to the WTC?
3. What new building methods and materials allowed the building of skyscrapers?
4. What did architects think about the future of skyscrapers after the WTC disaster?

Summary

Using your own words, give a summary of the text. It should be about 250 words long.

Vocabulary

1. Explain the following:
   a) steel supports (line 25)
   b) imploded (line 28)
   c) urbanisation (line 51)
   d) feedback (line 59)
   e) crush-proof (line 68)
2. Look in the text to find the nouns corresponding to these adjectives:
   a) safe
   b) accidental
   c) natural
   d) technological
   e) destructive
   f) hot

Translation

Translate lines 31 to 47.

Comment/Letter

Write a letter giving your opinion on the future of skyscrapers as a good place to live and work. Imagine the letter is addressed to the editor of the magazine which published this article.

Biotechnology develops the next generation of medicines

Do you prefer your potatoes fried, baked, mashed or in pill form? Under the lights in the research labs in Cambridge, England, a biotechnology company called Alpha Genetics has been growing genetically modified potatoes which are dried, crushed and served in capsules. And people will swallow these pills not as food, but as vaccines against such illnesses as hepatitis B, or cholera.
It is very important that new effective vaccines are found quickly as over 800,000 people die from hepatitis B alone every year. A problem with the current vaccines is that they must be refrigerated and injected by needles, a process that is not always easy or safe in the developing world.
So why not try potato pills? Nature has always been important in the medicine cabinet. Most people in the world still rely on traditional medicines based on plants and animals. And so many modern pharmaceutical drugs come from compounds discovered in nature. Very often the plants are found in areas which are ecologically sensitive, for example the fast disappearing tropical rainforests. This is one main reason why pharmaceutical companies are looking at plants and animals in a new way.
Through the progress of genetic research, scientists have been able to create plants which are genetically modified. Moreover, researchers are also using farm animals such as sheep or pigs to produce pharmaceutics in a process called "pharming". In this process animals are created that can produce human proteins. These proteins form the basis for medicines which can help people with various diseases. For example they can help people with genetic diseases and who are not able to produce these proteins for themselves.
But how is this achieved? Researchers concentrate on putting human genes into another mammal, which then can produce human protein in her milk. In Edinburgh, Scotland scientists produced Dolly, the first cloned sheep. However, Dolly was not a real success, because she could not be used to produce pharmaceutics. The reason was, although Dolly was a clone, she was not a transgenic animal. This means she did not have any implanted human genes and so could not produce the human proteins in her milk which are necessary to make vaccines.

The concepts of cloning and pharming have frightened a lot of people who now see cloning as a terrible reality which could have a negative effect on our world. Moreover, environmentalists are also sceptical about bioengineering. They fear that the alien genes introduced into genetically modified plants will escape into other agricultural or wild plants.
Nevertheless, biotechnology exists and has become big business. Research goes on at a rapid pace. For instance, an American company in Massachusetts made headlines by producing three goats that were both clones and transgenic. The researchers there hope to milk a protein from the goats and then use it in a drug which prevents blood clotting during heart surgery. This drug at the moment comes from human blood plasma which is a risk because the blood supply has sometimes been contaminated by such diseases as AIDS.
So where are we going? While companies like Alpha Genetics dream of vaccinating the world with genetically modified potato plants grown in a laboratory, other people fear that this is just the start of a science fiction reality which will affect our planet in a highly negative way.
The question which remains is "can we produce genetically modified plants and animals for the good of mankind without allowing them to escape and endanger our natural world?"

593 words

Exam preparation

Comprehension questions

■ After you have read the text carefully, answer the following questions in your own words as far as possible.

1. Why is the company using potatoes in its research?
2. Where do medicines often come from?
3. How does "pharming" work?
4. What have scientists in Massachusetts produced and why?
5. Why are most environmentalists often against genetic engineering?

Summary

■ Using your own words, give a summary of the text. It should be about 250 words long.

Vocabulary

1. Explain the following:
   a) vaccines (line 7)
   b) genetically modified plants (lines 54/55)
   c) clones (line 61)
   d) transgenic (line 61)
   e) bioengineering (line 53)
2. Find the nouns which correspond to the following verbs taken from the text:
   a) die (line 10)
   b) inject (line 13)
   c) serve (line 6)
   d) refrigerate (line 12)
   e) create (line 27)
   f) grow (line 4)
   g) prevent (line 63)
   h) vaccinate (line 70)

Translation

■ Translate lines 1 to 14.

Comment

■ Comment on the following statement, using your own knowledge and ideas.

"Cloning animals and using genetically modified plants has become a terrible and disastrous reality."

Future developments

The next twenty-five years

Nearly every day, somewhere in the world, scientific and technological developments are changing the dreams of science fiction into reality. February 14th 2002 was one such day when for the first time trading on the New York stock market was opened not by a human but by a humanoid robot. This two-legged robot looked like an astronaut and walked just like a human, even up and down slopes and stairs. It could change direction without having to stop and turn like other robots and it had posture control so that it didn't lose its balance and fall over like earlier robots had done. Although it was only 120 cm tall and weighed 52 kg, the robot was just the right size to be useful in society. Its height was the minimum necessary for it to do ordinary jobs like turning off lights or laying tables, but also meant that to humans it looked childlike and non-threatening.

And this was only the beginning. By the year 2020 scientists predict an explosion in technological developments which will change the world radically in a very short time. In two key technologies – computer power and DNA sequencing – we will see whole industries rise and fall as a result of unbelievable scientific advances.

By 2020 microprocessors will probably be as cheap and plentiful as waste paper is today and they will be found by the millions. This will mean that we will be able to place intelligent systems everywhere so that everything around us, including the way we communicate, work, play and live will change. This change will be possible because processors will be built into common objects in the home and workplace. Computer programs will be developed that can learn by trial and error. In this way, so called learning machines will be able to adjust their behaviour and intelligent software agents (knowbots or navigators) will routinely collect and filter information for users. Expert systems will be used regularly to help make decisions in management, medicine, engineering and other fields, while 'bio-chips' that store data in molecular bonds will also become available commercially.

Scientists also expect that the internet will wire up the entire planet and develop into a membrane consisting of millions of computer networks. In this way we will create an intelligent planet in which the internet will eventually give people access to all the wisdom of the human race.

By around 2020 however, both computer technology and DNA sequencing will run into difficulties. Because of the limits of silicon chip technology we will have to invent new technologies about which at the moment we know very little, from optical computers, molecular computers and DNA computers to quantum computers.

If we can find a solution to these difficulties in computer technology, then we may see an entirely new kind of technology come onto the market between 2020 and 2050. We may see for example today's walking robots develop into true robot automatons that have common sense, can understand human language, can recognize and manipulate objects in their environment and learn from their mistakes.

It is a development which some people may find exciting and others worrying. Whatever the case it is a development that will probably change our relationship with machines forever.

514 words

Exam preparation

Comprehension questions

■ After you have read the text carefully, answer the following questions in your own words as far as possible.

1. What new features did the robot which opened the stock market in 2002 have?
2. What will be the main changes in computer programs by the year 2020?
3. What effects will these changes have?
4. What developments may there be between 2020 and 2050 and how may people react to them?

Summary

■ Using your own words, give a summary of the text. It should be about 250 words long.

Vocabulary

1. Explain the following:
   a) humanoid robot (line 6)
   b) scientific advances (line 25)
   c) intelligent systems (line 29)
   d) by trial and error (line 35)
   e) internet (line 47)

2. Look in the text to find verbs for these nouns:
   a) weight
   b) prediction
   c) storage
   d) adjustment
   e) invention
   f) creation
   g) collection
   h) recognition

Translation

■ Translate lines 26 to 43.

Comment/Letter

It is the year 2050 and you are a journalist who has interviewed a producer of true robot automatons.

■ Write an article discussing the advantages and disadvantages of this new technology.

Abbildungsnachweis

Titelseite: Satellitenstation: MEV, Augsburg; Highway: Corbis (ML Sinibaldi), Düsseldorf
S. 9: oben, v. li. n. re.: MSI/BULLS; Corbis (Spiegel), Düsseldorf; Corbis (Woodcock; Reflections Photolibrary/CORBI), Düsseldorf; Mitte, v. li. n. re.: Siemens AG, Berlin; Corbis Digital Stock (RF/CORBIS), London; Siemens AG, Berlin; unten, v. li. n. re.: Corbis (Lewine), Düsseldorf; IBM Deutschland GmbH, Stuttgart; Getty Images (EyeWire), München
S. 10: www.britainonview.com, London
S. 11: Corbis (Ronnie Kaufman), Düsseldorf
S. 12: oben: MEV, Augsburg; unten: Elmar Feuerbach, Remshalden
S. 13: MEV, Augsburg
S. 14: Wolfram Büchel, Remshalden
S. 16: www.britainonview.com, London; Fabrikarbeiter: Corbis (Robert Essel NYC), Düsseldorf
S. 18: Getty Images, München
S. 19: Elmar Feuerbach, Remshalden
S. 20: IBM Deutschland GmbH, Stuttgart
S. 21: Mauritius, Mittenwald
S. 23: MEV, Augsburg; Frau mit Kind: Getty Images, München; älterer Mann: Henry Zeidler, Leipzig
S. 25: Alamy Images (RF), Oxfordshire
S. 26: MEV, Augsburg
S. 27: MEV, Augsburg
S. 28: MEV, Augsburg
S. 29: oben: Getty Images, München; unten: Corbis (Tim Page), Düsseldorf
S. 30: Corbis (Ben Wood), Düsseldorf
S. 31: MEV, Augsburg
S. 32: MEV, Augsburg; Krone: Getty Images (Ryan McVay), München; Parlament: www.britainonview.com, London
S. 33: oben li.: Corbis (Fleming), Düsseldorf; oben Mi.: IBM Deutschland GmbH (Edmondson), Stuttgart; Mitte: Siemens AG, Berlin; unten li.: Corbis (Edmondson), Düsseldorf; unten re.: Siemens AG, Berlin
S. 34: Corbis (C. Moore), Düsseldorf
S. 35: li.: IBM Deutschland GmbH, Stuttgart; Mi. u. re.: MEV, Augsburg
S. 36: IBM Deutschland GmbH, Stuttgart
S. 37: MEV, Augsburg
S. 38: Kessler-Medien, Saarbrücken
S. 40: oben: MEV, Augsburg; unten: Corbis (Randy Faris), Düsseldorf
S. 41: Corbis (Robert Landau), Düsseldorf
S. 42: Alamy Images (David Gould / Alamy), Oxfordshire
S. 43: Corbis (Roger Ressmeyer), Düsseldorf
S. 44: oben, v. li. n. re.: Corbis (FORESTIER YVES/CORBIS SYGMA), Düsseldorf; www.britainonview.com, London; MEV, Augsburg; www.britainonview.com, London; unten, v. li. n. re.: Corbis (Gabe Palmer), Düsseldorf; Corbis (LWA-Dann Tardif/CORBIS), Düsseldorf; Corbis (Richard T. Nowitz), Düsseldorf; Corbis (Benjamin Rondel), Düsseldorf
S. 45: Corbis (RF), Düsseldorf
S. 46: Pauline Ashworth, Stuttgart
S. 47: oben, v. li. n. re.: Getty Images, München; Corbis (Nick Hawkes), Düsseldorf; unten, v. li. n. re.: Alamy Images (David Crausby / Alamy), Oxfordshire; Siemens AG, Berlin; Corbis (Alan Schein Photography/CORBIS), Düsseldorf; TransportEnergy, London
S. 48: Highways Agency, London
S. 49: Getty Images, München
S. 51: v. oben n. unten: Corbis (Premium Stock/CORBIS), Düsseldorf; MEV, Augsburg; TransportEnergy, London; Corbis (Jeremy Horner), Düsseldorf
S. 53: Corbis (Annie Griffiths Belt), Düsseldorf
S. 54: BMW AG, München
S. 57: MEV, Augsburg
S. 58: Getty Images, München
S. 61: MEV, Augsburg
S. 62: oben: Getty Images (Glen Allison), München; unten li.: Corbis Digital Stock, London; unten re.: UTC Fuel Cells
S. 64: Fibrowatt Ltd, London
S. 65: oben: ullstein bild, Berlin; unten: Corbis (Ray Juno), Düsseldorf
S. 66: v. oben n. unten: Siemens AG, Berlin; IBM, Stuttgart; DaimlerChrysler AG, Stuttgart; MEV, Augsburg; IBM, Stuttgart
S. 67: Audi, Ingolstadt; Auspuff: Mauritius (Stock Image), Mittenwald
S. 68: Audi, Ingolstadt
S. 69: MEV, Augsburg
S. 71: Elmar Feuerbach, Remshalden
S. 73: oben, v. li. n. re.: Red Door VR, Leeds; Getty Images (B.C. Moller), München; Mitte, v. li. n. re.: The Homes of Football (Stuart Clarke), The Lake District; University of Leeds; Leeds Bradford International Airport, Leeds; unten, v. li. n. re.: reproduced courtesy of Leeds City Council; www.britainonview.com, London; Red Door VR, Leeds
S. 74: Klett-Archiv, Stuttgart
S. 75: www.graeme-peacock.com, Newcastle upon Tyne
S. 76: MEV, Augsburg
S. 77: oben: Getty Images RF, München; li.: MEV, Augsburg; re.: Creativ Collection Verlag GmbH, Freiburg; unten: Getty Images, München
S. 78: Corbis (Michel Setboun), Düsseldorf
S. 80: Corbis Digital Stock (RF/CORBIS), London
S. 81: Corbis (Bettmann/CORBIS), Düsseldorf
S. 82: li.: Getty Images (McVay), München; re.: Corbis (Strauss/Curtis/CORBIS), Düsseldorf
S. 83: Corbis Digital Stock (RF), London
S. 84: oben: Corbis (Dave G. Houser/CORBIS), Düsseldorf; unten: Corbis Digital Stock (RF), London
S. 85: oben: Getty Images RF (A Carpenter), München; unten: Corbis (Strauss/Curtis/CORBIS), Düsseldorf
S. 86: oben: Corbis (Bettmann/CORBIS), Düsseldorf; unten: Getty Images, München
S. 88: v. li. n. re.: Geoatlas, Hendaye; Corbis (David H. Wells/CORBIS), Düsseldorf; Steve Skjold / Alamy; Corbis (Randy Duchaine/CORBIS), Düsseldorf
S. 89: oben, v. li. n. re.: Corbis (Lightscapes Photography, Inc.), Düsseldorf; Corbis Digital Stock (RF), London; Corbis (Pete Saloutos), Düsseldorf; Mitte, v. li. n. re.: Corbis (Michael Maconachie), Düsseldorf; E-Lance Media GmbH, Gauting; unten: Corbis (Ron Watts), Düsseldorf; Satellitenstation: MEV, Augsburg
S. 90: IBM, Stuttgart
S. 94: Corbis (Michael Maconachie), Düsseldorf
S. 98: Corbis (Lightscapes Photography, Inc.), Düsseldorf
S. 101: oben: Corbis (Mug Shots), Düsseldorf; unten: MEV, Augsburg
S. 102: IBM Deutschland GmbH, Stuttgart
S. 105: IBM Deutschland GmbH, Stuttgart; Scanner u. Drucker: Hewlett-Packard, Böblingen
S. 106: Corbis Digital Stock (RF), London
S. 112: MEV, Augsburg
S. 114: Corbis (Ron Watts), Düsseldorf
S. 116: Corbis (Pete Saloutos), Düsseldorf
S. 118: E-Lance Media GmbH, Gauting
S. 199: Klett-Archiv, Stuttgart
S. 200: Klett-Archiv, Stuttgart

In einigen Fällen ist es uns trotz intensiver Bemühungen nicht gelungen, die Rechteinhaber zu ermitteln. Wir bitten diese, sich mit dem Verlag in Verbindung zu setzen.

Indefinite article (Unbestimmter Artikel)

Form

a	college student university	vor **gesprochenen** Konsonanten
an	interview address hour	vor **gesprochenen** Vokalen

Verwendung

I have bought **a** book for you.	für zählbare Begriffe in der Einzahl
He is **a** teacher.	vor Berufsbezeichnungen

Definite article (Bestimmter Artikel)

Form

the	course language European	vor **gesprochenen** Konsonanten Aussprache [ðɛ]
the	engineer office hour	vor **gesprochenen** Vokalen Aussprache [ðiː]

Verwendung

the people of London **the** life of Lucy Jordan **the** University of Manchester	bei Substantiven (Personen, Dingen, Abstrakta), wenn sie näher bestimmt sind oder werden
the Americans	für die Gesamtheit von Personen
the United States	für Ländernamen im Plural
the Thames, **the** North Sea	für Flüsse und Meere
the Browns	für Eigennamen im Plural
On their tour they visited **the** church.	für die konkrete Bedeutung als Gebäude

Verwendung ohne Artikel

Der Artikel wird abweichend vom Deutschen **nicht** verwendet:	
nature, life, time, people	für abstrakte Begriffe und Sammelbezeichnungen, wenn sie nicht näher bestimmt sind oder werden
Trafalgar Square, Buckingham Palace	für Ortsbezeichnungen
Switzerland, Normandy, Lake Windermere	für Länder, Berge und Seen im Singular
church, school, college When does church begin?	für Gebäudebezeichnungen, wenn die normale Nutzung gemeint ist
Americans	für einzelne Angehörige einer Personengruppe

Plural (Mehrzahl)

Form

student college	student**s** college**s**	Die meisten Nomen bilden den Plural durch Anhängen von -s.
universit**y** cit**y**	universit**ies** cit**ies**	Konsonant + y wird zu -ies.
addre**ss** mat**ch**	address**es** match**es**	Nach Zischlauten (z. B. s, ss, ch, sh) wird -es angehängt.
man woman child life	**men** **women** **children** **lives**	Einige Nomen haben unregelmäßige Pluralformen.

Verwendung

information, weather, nature, knowledge, work, furniture	Einige Nomen treten nur im Singular auf.
trousers, jeans, glasses, clothes, thanks	Einige Nomen treten nur im Plural auf.
news, the United States, the United Nations, economics, mathematics, electronics ! No news **is** good news.	Einige Nomen treten nur in der Pluralform auf, das Verb steht jedoch im Singular.

Some important prepositions

Zeitliche Präpositionen

on 20th September **on** Saturday **on** weekdays	**at** six o'clock **at** the age of 16 **at** night **at** noon **at** midnight	**by** day
		twelve minutes **to** five
in 1992 **in** the morning **in** the evening		a quarter **past** six

Präpositionen der Art und Weise

go **by** bus, car, bike
go **on** foot

Räumliche Präpositionen

go **to** the cinema go **to** college I've been **to** Scotland
at home **at** school **at** the supermarket

Possessive case (Genitiv)

s'-Genitiv

Verwendung bei Personen, Tieren, Zeiten

Henry**'s** exam the baker**'s** (shop)	Nomen im Singular + **'s**
his parents' computer shop the Greens' house	Nomen im Plural + **'**
people**'s** faces children**'s** books	bei Nomen im Plural **ohne** s Anhängung von **'s**
today**'s** newspaper two weeks' pay	Bei Zeiten gelten die gleichen Regeln.

of-Genitiv

Verwendung bei Sachen

the title **of** the book the price **of** the computers	Der *of*-Genitiv wird im Singular und im Plural verwendet.

Comparison of adjectives (Steigerung von Adjektiven)

Form

cheap small	cheap**er** small**er**	cheap**est** small**est**	Steigerung der kurzen Adjektive mit -er und -est
nice	nic**er**	nic**est**	stummes -e entfällt
heavy	heav**ier**	heav**iest**	-y wird zu -ier und -iest
big	bi**gg**er	bi**gg**est	Nach kurzen Vokalen wird der Endkonsonant verdoppelt.
good bad little	**better** **worse** **less**	**best** **worst** **least**	Einige Adjektive haben unregelmäßige Steigerungsformen.
careful difficult	**more** careful **more** difficult	**most** careful **most** difficult	Alle dreisilbigen und mehrsilbigen Adjektive und alle zweisilbigen Adjektive, die nicht auf -er, -le, -y, -ow enden, werden mit *more* und *most* gesteigert.
careful difficult	**less** careful **less** difficult	**least** careful **least** difficult	Eine Verminderung der Adjektive ist mit *less* und *least* möglich.

Verwendung

In town centres walking is often **as fast as** going by car.	Gleichheiten werden mit *as … as* ausgedrückt.
Travelling by plane is **faster than** travelling by train. Travelling by car is **more expensive than** walking.	Komparative werden mit … *than* gebildet.
A bicycle is **not so expensive as** (isn't as expensive as) a car. A bicycle is **less expensive than** a car.	Eine andere Form Ungleichheit auszudrücken, ist *not so … as* oder umgangssprachlich *not as … as*. Wie im Deutschen kann man auch „weniger als" verwenden.
The Concorde was **the fastest** airliner from Paris to New York.	Der Superlativ wird nur benutzt, wenn mehr als zwei Dinge miteinander verglichen werden.

Adjective and adverb (Adjektiv und Adverb)

Form

Adjective	Adverb	
cheap lazy horrible economic fast good	cheap**ly** laz**ily** horrib**ly** econom**ically** **fast** **well**	Das Adverb wird gebildet aus Adjektiv + -ly. -y wird zu -ily -le wird zu -ly -ic wird zu -ically Einige Adverbien bilden unregelmäßige Formen.

Adjective	Adverb	Adverb	
! late	**late** (spät)	**lately** (in letzter Zeit)	Einige Adverbien bilden mehrere Formen.
hard	**hard** (hart)	**hardly** (kaum)	
near	**near** (nah)	**nearly** (beinahe)	

Verwendung

There are many **new computers** on the market. **They** are still **expensive**.	Adjektive beziehen sich auf Nomen oder Pronomen.
We can't **control** the technology **exactly**. Comics are **really popular** with children. Modern computers work **extremely quickly**. **Fortunately, she e-mailed the information to me in time.**	Adverbien beziehen sich auf – Verben – Adjektive – Adverbien – ganze Sätze
! He **felt nervous** about his new job.	Nach Verben, die einen Zustand darstellen (z. B. *feel, be, become, get, seem, appear, keep*) folgt ein Adjektiv.

Word order with adverbs and adverbials (Wortstellung bei Adverbien und adverbialen Bestimmungen)

Die Stellung der Adverbien und adverbialen Bestimmungen (a. B.) ist im Englischen sehr flexibel. Folgende Stellungen sind allerdings üblich:	
At nine o'clock he caught the bus **to work**.	Adverbien/a. B. der Zeit stehen am Anfang oder Ende des Satzes. Adverbien/a. B. des Ortes stehen in der Regel am Ende des Satzes.
Mr Woods **always** travels to work by train. The train arrives **punctually**.	Adverbien/a.B. der Häufigkeit stehen vor dem (Haupt-)Verb. Adverbien/a. B. der Art und Weise stehen in der Regel nach dem Verb (den Verben).
The bus drove **slowly** through **London during the rush hour**.	Treten mehrere Adverbien/a.B. zur gleichen Zeit auf, gilt die Regel: Art und Weise vor Ort und Zeit.

Some or any?

Verwendung

I need **some** help with this program. **Some** newspapers in Britain only appear on Sundays.	in Aussagesätzen vor unbestimmten Mengen (etwas) oder Anzahlen (einige)
They didn't make **any** money with the new campaign. Some TV stations do not show **any** adverts.	in verneinten Sätzen vor unbestimmten Mengen oder Anzahlen (kein oder keine)
Have they done **any** research on the new product? Have you got **any** ideas for a new logo?	in Fragesätzen vor unbestimmten Mengen (etwas) oder Anzahlen (welche?)
Would you like me to show you **some** examples?	in Fragesätzen, wenn sie eine Aufforderung, Bitte oder ein Angebot enthalten
We never needed **any** electronic devices in the past.	nach Wörtern mit negativer Bedeutung

Some/any + -one/-body/-thing/-where

Somebody wants to see you. We haven't been **anywhere** this weekend. Have you got **anything** to say?	Die Regeln zum Gebrauch von *some* und *any* gelten auch für diese Formen.

Much – many – a lot of

Form

much **a lot of**	viel	nicht zählbar
many **a lot of**	viele	zählbar

Verwendung

He spends **a lot of** time practising his communication skills. **A lot of** (many) people have computers in their homes.	In Aussagesätzen wird in der Regel *a lot of* bevorzugt.
I don't have **a lot of/much** money to spend on computers. There aren't **a lot of/many** experts on the staff.	In verneinten Sätzen kann man beide Formen gebrauchen.
How **much** money do you earn? How **many** adverts did they show?	Nach *how* müssen *much* oder *many* stehen.

Little/a little – few/a few

Form

We had **little** help during our training. John had only **a little** time to explain.	wenig ein bisschen	unzählbar
Few computers never break down. There are **a few** good adverts on TV.	wenige ein paar	zählbar

Modal auxiliaries (Modale Hilfsverben)

Form

Modale Form	Ersatzform	Bedeutung
must	**have to**	müssen
can	**be able to**	können
can	**be allowed to**	dürfen
mustn't **can't**	**not be allowed to**	nicht dürfen
can't	**not be able to**	nicht können
needn't	**not have to**	nicht brauchen/nicht müssen

Verwendung

Cars **must/have to** stop at red lights. You **needn't/don't have to** pay in that car park. John **can't/isn't able to** find a parking space.	In der Gegenwart kann entweder die modale oder die Ersatzform benutzt werden.
We **had to** turn right at the traffic lights. He **didn't have to** wait as the lights were green. Jane **couldn't/wasn't allowed to** leave her car outside the house, as there was a "No Parking" sign.	Mit Ausnahme von *could* oder *couldn't* können die modalen Hilfsverben keine anderen Zeiten bilden. Für die anderen Zeiten werden die Ersatzformen verwendet.

Defining relative clauses (Notwendige Relativsätze)

Verwendung

Ein Relativsatz ist notwendig, wenn die darin enthaltene Information nötig ist, um zu erkennen, welche Person oder Sache gemeint ist.	
The man **who/that** is getting on the bus is on his way to the office.	Bei Personen werden die Relativpronomen *who* oder *that* verwendet.
The bus **which/that** is waiting at the traffic lights is going to London.	Bei Sachen wird *which* oder *that* verwendet.
He greeted the man **(who)** he saw on the platform. He read the paper **(which)** he had bought that morning.	Wenn das Relativpronomen Objekt des Satzes ist, kann es weggelassen werden.

Non-defining relative clauses (Nicht notwendige Relativsätze)

Verwendung

Ein Relativsatz ist nicht notwendig, wenn die darin enthaltene Information nicht nötig ist, um zu erkennen, welche Person oder Sache gemeint ist.	
Henry's wife, **who is 58**, was waiting for him at the station.	Ein Relativsatz wird durch Kommas vom Hauptsatz getrennt.
The train, **which was an Intercity**, left punctually.	*that* darf hier nicht verwendet werden.
Henry, **who I met at the station**, is retiring next week.	Wenn das Relativpronomen Objekt eines Relativsatzes ist, darf es nicht weggelassen werden.

The Tenses (Die Zeiten)

Present simple

Form

I/you/we/they he/she/it	**live** **lives**	in Leeds.	Aussage
I/you/we/they he/she/it	**do not (don't) live** **does not (doesn't) live**	in Birmingham.	Verneinung

Do **Does**	I/you/we/they he/she/it	**(not)**	**live**	in Liverpool?	(verneinte) Frage
Where **Why**	**do** you **does** she	**live**? **smoke**?			

!

Who **How many students**	**likes** **live**	basketball? at home?	Ist das Fragewort Subjekt oder Teil des Subjekts, wird *do/does* nicht verwendet.

Besonderheiten bei der Schreibung der s-Endung

go, do watch, kiss carry, try	go**es**, do**es** watch**es**, kiss**es** carr**ies**, tr**ies**

Verwendung

She **has** a brother.	bei einem Dauerzustand
She **(usually) goes** to college by bus. He **(often) helps** at home.	bei regelmäßigen oder wiederholten Handlungen (oft mit Häufigkeitsadverbien wie *usually, normally, often, sometimes, never, always*)
She **doesn't smoke**.	bei Gewohnheiten
He **has** a flat in Mayfield. I **think** that this **is** a good idea.	bei bestimmten Verben, wenn sie einen Zustand beschreiben, z.B. *be, have, look, think, see, know, like, want*
The train **arrives** at 8 o'clock tomorrow morning.	bei Fahrplänen und Veranstaltungsprogrammen (mit Zukunftsbezug)

Present continuous

Form

I you/we/they he/she/it	**am** **are** **is**	**(not)**	**reading** a book.	Aussage und Verneinung

Kurzform der Verneinung: aren't, isn't

Am **Are** **Is**	I you/we/they he/she/it	**(not)**	**reading** a magazine?	Frage und verneinte Frage

Besonderheiten bei der Schreibung

come, take sit, run	com**ing**, tak**ing** si**tt**ing, ru**nn**ing

Verwendung

They **are talking** about their courses.	bei Handlungen, die im Moment des Sprechens stattfinden
She **is studying** for her exam this year.	bei Handlungen, die vorübergehend stattfinden, nicht aber unbedingt im Augenblick
She **is having** a good time. A lot of people **are thinking** about their future.	bei bestimmten Verben (siehe *present simple*), wenn sie eine vorübergehende Handlung und keinen Zustand beschreiben
I **am meeting** him tomorrow.	bei zukünftigen Handlungen (mit Zeitbestimmung)
Some students **are always complaining**.	zur gefühlsbetonten Darstellung von Handlungen in Verbindung mit *always*

Past simple

Form

She You	work**ed** need**ed**	in Glasgow a new car.	last year.	Aussagen Grundform + -ed (Infinitiv)

Besonderheiten

mov**e** tr**y** sto**p** go meet	mov**ed** tr**ied** sto**pped** **went** **met**	stummes -e entfällt Konsonant + y wird zu ie Lautverdopplung unregelmäßige Verben haben besondere Formen für das *simple past* dabei wird die 2. Form verwendet

He We	**did not** **(didn't)**	**need** **lose**	special software. the money.	Verneinung

Did	you they it	**(not)**	**move** **live** **take**	to Glasgow? in Birmingham? more than 6 months?	(verneinte) Frage
Why did	he	**(not)**	**stop**	in Liverpool?	

Verwendung

Jane **moved** to Glasgow 7 years ago. First she **worked** for a computer company.	bei Handlungen zu einem bestimmten Zeitpunkt oder in einem abgeschlossenen Zeitraum in der Vergangenheit (oft mit einer Zeitbestimmung wie z. B. *yesterday, last year, in 1988, … ago, from … until*)

Past continuous

Form

I/He/She/It	**was**	**(not)** **(wasn't)**	**standing**	on the platform.	(verneinte) Aussage
We/You/They	**were**	**(not)** **(weren't)**			

Verwendung

	As/while he **was driving** to the airport, it started to rain. A lot of passengers **were waiting** at the counter when Jason went into the bank.	für eine Handlung, die schon im Gange war, als eine neue Handlung eintrat
!	While I **was reading** a book, my friend **was writing** a letter.	Bei gleichzeitigem Verlauf zweier Handlungen steht die Verlaufsform in beiden Fällen.
!	What **did** you **do** when Susan **came** in? When she **came** in I **put** down my newspaper and talked to her.	Für Handlungen, die nacheinander stattfanden, wird in beiden Fällen *past simple* benutzt

Present perfect simple

Form

I/You/We/They	**have**	**(not)** **(haven't)**	**had** a car for three years.	(verneinte) Aussage *have/has* + 3. Form
He/She/It	**has**	**(not)** **(hasn't)**	**sold** 20 million CDs. **been** to Scotland.	

Have	I/you/we/they	**(not)**	**made** a loss? **taken** a photo? **been** here?	(verneinte) Frage
Has	he/she/it			

Verwendung

I **have repaired** the car.	bei Handlungen in der Vergangenheit ohne Zeitangabe – das Ergebnis ist oft wichtiger als die Zeitangabe
They **have sold** 50,000 cars **up till now**.	mit Zeitbestimmungen, die einen Zeitraum beschreiben, der noch andauert, z. B. *today, this week, so far, in the last ten years*
He has **already paid** the bill.	mit bestimmten Adverbien wie z. B. *ever, never, always, yet, just, already*
How long have you **known** about this book? The Browns **have lived** there **for** 3 years. We **have lived** here **since** 1986.	bei nicht abgeschlossenen Zuständen mit *how long, since* und *for* – *since* bezieht sich dabei auf einen Zeitpunkt, *for* auf einen Zeitraum

Present perfect continuous

Form

I/you/we/they	**have**	**been standing.**	Present simple von *have* + *been* + -ing-Form

Verwendung

How long have you been waiting? I've been waiting – for half an hour. – since ten o'clock.	bei Handlungen, die in der Vergangenheit begannen und noch andauern – meist in Verbindung mit *how long, since, for*

Past perfect simple

Form

I/You/He/She/ It/We/They	**had**	**(not)** **(hadn't)**	**travelled** to Paris.	(verneinte) Aussage

Verwendung

After he **had arrived** at the hotel, he went to his room. When I arrived at the station, the ticket office **had** already **closed**.	für Handlungen oder Zustände, die vor einem Zeitpunkt in der Vergangenheit abgeschlossen waren

Future with 'going to'

Form

I He/She/It We/You/They	**am ('m)** **is ('s)** **are ('re)**	**(not)**	**going to** watch TV.	(verneinte) Aussage

Am **Is** **Are**	I he/she/it we/you/they	**going to have**	a cup of coffee?	Frage

Verwendung

She **is going to answer** the phone.	bei Absichten und Vorhaben
That fax machine **is going to break** down.	bei Vorhersagen, die aufgrund bereits bekannter Fakten oder bisheriger Erfahrungen sicher oder logischerweise in Erfüllung gehen müssen

Future with 'will'

Form

I/You He/She/It We/They	**will** **will not (won't)**	**use** alternative energy. **waste** water.	Aussage Verneinung

Will	I/you he/she/it we/they	**use** talking computers?	Frage

Verwendung

Computers **will probably do** most of the work.	bei Vorhersagen mit */suppose/expect* und *probably*
Just a minute, **I'll help** you with your shopping.	bei Entscheidungen, die im Moment des Sprechens getroffen werden
I'll phone you tomorrow, I promise.	bei Versprechen
Will you **make** some coffee, please.	um eine Bitte auszudrücken

Question tags (Frageanhängsel)

Form

Aussagesatz	Frageanhängsel	
You can drive, Pollution is increasing, Trams will improve our public transport service,	**can't you?** **isn't it?** **won't they?**	Das Hilfsverb wird in dem Frageanhängsel wiederholt. Ist der Aussagesatz positiv, so wird das Frageanhängsel verneint.
Statistics aren't always correct, You mustn't drive over 70 mph on motorways,	**are they?** **must you?**	Ist der Aussagesatz verneint, so wird das Frageanhängsel bejaht.
We enjoyed the journey, She usually goes to Britain,	**didn't we?** **doesn't she?**	Wenn der Aussagesatz kein Hilfsverb enthält, wird das Frageanhängsel mit *to do* in der Zeit des Aussagesatzes gebildet.

Verwendung

	Frageanhängsel werden verwendet:
Air travel is really comfortable, **isn't it?**	beim Wunsch nach Bestätigung
You didn't walk here, **did you?**	bei Überraschung
Oh, it's not another traffic jam, **is it?**	bei Verärgerung
Well, that's a really new idea, **isn't it?**	bei ironischen Aussagen

Reported speech (Indirekte Rede)

Form

Aussagen

	Die Zeitformen ändern sich:
"We are protesting about pollution." Harry said they **were protesting** about pollution. "He doesn't buy cans." June told me he **didn't buy** cans.	*Present tenses* werden zu *past tenses*.
"Watsons opened four supermarkets in 1995." Fred pointed out that Watsons **had opened** four supermarkets in 1995. "They have started a bus service." He added that they **had started** a bus service.	*Simple past* und *present perfect* werden zu *past perfect*.
"There will be more 'green' products." Sue went on to say that there **would be** more 'green' products.	*will* wird zu *would*.
"I can save more energy." Harry told me that **he** could save more energy.	Die Pronomen werden angepasst.
"We are meeting the town planner tomorrow." Sue added that they were meeting the town planner **the next day**. "I bought the car three days ago." He said that he had bought the car **three days before**.	Orts- (z. B. *here* wird zu *there*) und Zeitbestimmungen werden angepasst.

Fragen

"What are you doing about waste?" **They asked what we were doing** about waste. "When did you introduce bicycle routes?" **We wanted to know when they had introduced** bicycle routes.	Bei Fragen wird die Wortstellung geändert.

Verwendung

Indirekte Rede wird benutzt, wenn berichtet werden soll, was jemand gesagt oder gefragt hat.

Conditionals (Bedingungssätze)

Form

Type 1

If-Satz	Hauptsatz	
If **we fit** a shower,	**we will** save water.	*Present simple* wird für das Verb im If-Satz, Futur mit *will* für das Verb im Hauptsatz verwendet.

Type 2

If-Satz	Hauptsatz	
If the government **developed** solar energy,	there **would be** less pollution.	*Past simple* wird für das Verb im If-Satz, *Conditional I* für das Verb im Hauptsatz verwendet.
If I **were** Environment Minister,	I **could** put my ideas about energy into practice.	*could* oder *might* sind Alternativen zu *would* im Hauptsatz.

Type 3

If-Satz	Hauptsatz	
If Tom and Kate **had insulated** their roof years ago,	they **would have saved** a lot of money.	*Past perfect* wird für das Verb im If-Satz, *Conditional II* für das Verb im Hauptsatz verwendet.

Verwendung

Type 1 wird gebraucht, wenn die Voraussetzungen für die Erfüllung der Bedingung schon existieren und es deswegen wahrscheinlich ist, dass die Handlung im If-Satz in Erfüllung gehen wird (wir planen, eine Dusche einzubauen).
Type 2 wird gebraucht, wenn a) wir nicht erwarten, dass die Handlung im If-Satz eintreten wird. Wir stellen uns die Situation rein theoretisch vor. b) die Handlung im If-Satz gar nicht eintreten kann, weil sie in Gegensatz zu den Tatsachen steht (ich bin nicht der Umweltminister).
Type 3 wird gebraucht, wenn die Handlung im If-Satz nicht eintreten kann, weil sie in der Vergangenheit liegt. Es kann an der Situation nichts mehr geändert werden.

Passive voice (Passiv)

Form

The robots **are equipped** with sensors. This motor **was not made** in Great Britain. **Will** the new car **be sold** in Japan?	Form von *be* in der jeweiligen Zeit + Partizip Perfekt (3. Form) des Verbs
Now the wheels **can be fitted**. The car **must be tested** first.	auch mit Hilfsverben

Verwendung

<table>
<tr><td colspan="2">These cars were made in Germany.
The handbook is written in English.</td><td>hauptsächlich in der Schriftsprache, wenn der Ausführende unbekannt, unwichtig oder selbstverständlich ist</td></tr>
<tr><td colspan="2">This information is checked by the central computer.</td><td>Wenn der Ausführende allerdings genannt werden soll, dann benutzt man die Präposition by (by-agent).</td></tr>
<tr><td>It is said
It is reported
It is believed
It is supposed</td><td>that computers will become even more important in the future.</td><td rowspan="2">bei Verben des Meinens und Berichtens</td></tr>
<tr><td colspan="2">He is said to be rich.</td></tr>
<tr><td colspan="2">English is spoken here. = Man spricht Englisch.
We will be told the result later = Man wird uns später das Ergebnis sagen.</td><td>im Deutschen wird ein Aktivsatz mit ‚man' bevorzugt</td></tr>
</table>

Infinitive (Infinitiv)

Verwendung

ohne *to*

We **must write** a report about our trip to England.	nach den meisten Hilfsverben
Our new teachers **make us work** very hard, but on Friday they **let us go** home early.	nach *make* und *let* + direktem Objekt

mit *to*

The teacher **asked me to show** my photos of Newcastle. She **told them to wait** outside for a moment.	nach bestimmten Verben (*ask/tell/advise/expect* etc.)
This is not **easy to understand**. I am **surprised to hear** that.	nach Adjektiven
I do not know **what to do**. I soon found out **where to go**.	nach Fragewörtern

Gerund (Gerundium)

Verwendung

Drinking and **driving** is dangerous.	als Subjekt
I **enjoy swimming.**	als Objekt nach Verben ohne Präpositionen, wenn eine allgemeingültige Situation beschrieben wird: z. B. *enjoy/like/dislike/hate/stop/start/avoid/suggest/mind/love/recommend/prefer*
I **look forward to seeing** you again. He is **tired of waiting** here. They saw the **danger of destroying** the environment.	nach Verben/Adjektiven/Substantiven mit Präpositionen
He has helped me a lot **by giving** me that map of Newcastle. The child crossed the road **without looking**.	nach Präpositionen mit adverbialer Bedeutung, z. B. *by/without/instead of*
! Normally I enjoy swimming but today I **would prefer to play** tennis. I remember learning those rules. = Ich erinnere mich daran, dass ich die Regeln gelernt habe. I must remember to learn these rules. = Ich muss daran denken, diese Regeln zu lernen.	aber kein Gerundium, wenn es sich um eine Ausnahmesituation handelt (oft mit *would*) oder wenn die Bedeutung es nicht zulässt, z. B. bei *stop/start/remember*

Participle (Partizip)

Form

Das Partizip besteht aus zwei Formen

waiting/going/watching	Partizip Präsens
waited/gone/watched	Partizip Perfekt

Verwendung

The weather forecast for the **coming** week … The weather in New York last week was rather **mixed**.	als Adjektiv
You can find several families **living** in one flat. There are a lot of restaurants there especially **built** for the tourists.	anstatt von Relativsätzen
After talking to the press they started their demonstration. **Having discussed** the problem with the President personally, they flew home to California.	als Verkürzung von Adverbialsätzen (*after/before/because/while* etc.)

Unit vocabulary

AE = American English
BE = British English

Unit 1

Starter

BTEC National Diploma [ˌbiːtiːiːˈsiː ˌnæʃənl dɪˈpləʊmə]	etwa: Fachhochschulreife-prüfung
business studies [ˈbɪznɪs‿ˌstʌdɪz]	etwa: Fachbereich Wirtschaft
certificate [səˈtɪfɪkət]	Zeugnis
comprehensive school [kɒmprɪˈhensɪv ˌskuːl]	Gesamtschule
engineering studies [ˌendʒɪˈnɪərɪŋ ˌstʌdɪz]	etwa: Fachbereich Technik
GCSE [ˌdʒiːsiːesˈiː] = General Certificate of Secondary Education [ˌdʒenərəl səˈtɪfɪkət əʌ ˌsekəndrɪ‿ˌedʒʊˈkeɪʃn]	etwa: Fachoberschulreife
public service [ˌpʌblɪk ˈsɜːvɪs]	öffentlicher Dienst
social studies [ˈsəʊʃl ˌstʌdɪz]	Sozialwissenschaften
studies [ˈstʌdɪz]	Kurs, Studium

A1

accountancy [əˈkaʊntənsɪ]	Rechnungswesen
application [ˌæplɪˈkeɪʃn]	Anwendung
bus pass [ˈbʌs ˌpɑːs]	Busausweis
CAD [ˌsiːeɪˈdiː] (Computer Aided Design) [kəmˈpjuːtər‿ˌeɪdɪd‿dɪˈzaɪn]	Computerunterstützte Zeichnung
CAM [ˌsiːeɪˈem] (Computer Aided Manufacturing) [kəmˈpjuːtər‿ˌeɪdɪd ˌmænjəˈfæktʃərɪŋ]	Computerunterstützte Fertigung
civil engineering [ˈsɪvl‿ˌendʒɪˈnɪərɪŋ]	Bauwesen
control [kənˈtrəʊl]	Kontrolle, hier: Regeltechnik
electrical engineering [ɪˈlektrɪkl‿ˌendʒɪˈnɪərɪŋ]	Elektrotechnik
equipment [ɪˈkwɪpmənt]	Ausrüstung
finance [ˈfaɪnæns]	Finanzwesen
full-time [ˌfʊl ˈtaɪm]	Vollzeit
general science [ˌdʒenərəl ˈsaɪəns]	allgemeine Naturwissenschaften
grant [grɑːnt]	Zuschuss, Bafög
health care [ˈhelθ ˌkeə]	Gesundheitsfürsorge
industrial [ɪnˈdʌstrɪəl]	industriell
instrumentation [ˌɪnstrəmenˈteɪʃn]	Messtechnik
level [ˈlevl]	Stufe, Niveau
mechanical engineering [məˈkænɪkl‿ˌendʒɪˈnɪərɪŋ]	Maschinenbau
office skills [ˈɒfɪs‿ˌskɪlz]	Bürofertigkeiten
paragraph [ˈpærəgrɑːf]	Absatz
part-time [ˌpɑːt‿ˈtaɪm]	Teilzeit
principle [ˈprɪnsəpl]	Grundsatz, Prinzip
psychology [saɪˈkɒlədʒɪ]	Psychologie
qualification [ˌkwɒlɪfɪˈkeɪʃn]	Qualifikation, Befähigung
to run courses [rʌn ˈkɔːsɪz]	Kurse anbieten
sociology [ˌsəʊʃɪˈɒlədʒɪ]	Soziologie
statistics [stəˈtɪstɪks]	Statistik

A2

bedsitter [bedˈsɪtə]	Einzimmerwohnung
beginner [bɪˈgɪnə]	Anfänger(in)
campus [ˈkæmpəs]	Campus, Universitätsgelände
coach [kəʊtʃ]	Trainer(in)
dialogue [ˈdaɪəlɒg]	Dialog
female [ˈfiːmeɪl]	weiblich
formula [ˈfɔːmjələ]	Formel
information sheet [ˌɪnfəˈmeɪʃn ˌʃiːt]	Informationsblatt
long time no see [lɒŋ ˌtaɪm nəʊ ˈsiː]	lange nicht gesehen
male [meɪl]	männlich
packing [ˈpækɪŋ]	Verpackung
pocket money [ˈpɒkɪt ˌmʌnɪ]	Taschengeld
to rush [rʌʃ]	eilen
social studies [ˈsəʊʃl ˌstʌdɪz]	Sozialwissenschaften

C1

adventure [ədˈventʃə]	Abenteuer
as well as [əz‿ˈwel‿əz]	sowie
to be into [biː‿ˈɪntuː]	Fan von etwas sein
biscuit [ˈbɪskɪt]	Keks
bold type [ˌbəʊld‿ˈtaɪp]	Fettdruck
to broaden the mind [ˌbrɔːdn ðə ˈmaɪnd]	den Horizont erweitern
comment [ˈkɒment]	Bemerkung, Kommentar
to cope [kəʊp] (with)	mit etwas klarkommen
district [ˈdɪstrɪkt]	Bezirk, Gegend
exchange [ɪksˈtʃeɪndʒ]	Austausch
to get on like a house on fire [get‿ˈɒn laɪk‿əˌhaʊs‿ɒn ˈfaɪə]	sich sehr gut verstehen

to get to know [get‿tʊ 'nəʊ]	kennen lernen
head of department [ˌhed‿əv dɪ'pɑːtmənt]	hier: Fachbereichsleiter
heavy metal [ˌhevɪ 'metl]	Heavy Metal (harte Rockmusik)
host [həʊst]	Gastgeber(in)
joint [dʒɔɪnt]	gemeinsam
metal ['metl]	Metall
moor [mɔːr]	Moor
organizer ['ɔːgənaɪzə]	Veranstalter(in)
project ['prɒdʒekt]	Projekt
to put (s.o./sth.) to the test [ˌpʊt‿tʊ ðə 'test]	(jmdn./etwas) auf die Probe stellen
remark [rɪ'mɑːk]	Bemerkung
sights [saɪts]	Sehenswürdigkeiten
smoker ['sməʊkə]	Raucher(in)
staffroom ['stɑːfruːm]	Lehrerzimmer
workshop ['wɜːkʃɒp]	Werkstatt

C2

can't stand [kɑːnt‿'stænd]	nicht ausstehen können
informal [ɪn'fɔːml]	informell
to look like ['lʊk laɪk]	aussehen wie
personality [ˌpɜːsə'nælətɪ]	Persönlichkeit
scheme [skiːm]	Projekt, Plan, Programm
Yours [jɔːz]	Dein(e)

D1

a bit [ə bɪt]	ein bisschen
enthusiastic [ɪnˌθjuːzɪ'æstɪk]	begeistert
for instance [fər‿'ɪnstəns]	beispielsweise
hard-working [ˌhɑːd'wɜːkɪŋ]	fleißig
in my view [ɪn‿'maɪ ˌvjuː]	meiner Ansicht nach
locals ['ləʊklz]	Einwohner
phrase [freɪz]	Ausdruck

Unit 2

Starter

active ['æktɪv]	aktiv
advertising manager ['ædvətaɪzɪŋ ˌmænɪdʒə]	Werbeleiter(in)
creative [kriː'eɪtɪv]	kreativ
dancer ['dɑːnsə]	Tänzer(in)
to discover [dɪ'skʌvə]	entdecken
flight attendant ['flaɪt‿əˌtendənt]	Flugbegleiter(in)
hairdresser ['heəˌdresə]	Friseur(in)
independent [ˌɪndɪ'pendənt]	unabhängig
logical ['lɒdʒɪkl]	logisch
mark [mɑːk]	Zensur, Note
novel ['nɒvl]	Roman
office worker ['ɒfɪs ˌwɜːkə]	Büroangestellte(r)
officer ['ɒfɪsə]	Beamte(r), Offizier(in)
on your own [ˌɒn‿jər‿'əʊn]	allein
per [pə]	pro
practical ['præktɪkl]	praktisch
puzzle ['pʌzl]	Rätsel
salesman/saleswoman ['seɪlzmən /'seɪlzˌwʊmən]	Verkäufer(in)
singer ['sɪŋə]	Sänger(in)
specialist ['speʃəlɪst]	Spezialist(in)
sportsman ['spɔːtsmən]	Sportler
sportswoman ['spɔːtsˌwʊmən]	Sportlerin
suggestion [sə'dʒestʃn]	Vorschlag
suitable ['sjuːtəbl]	geeignet
tourist guide ['tʊərɪst ˌgaɪd]	Fremdenführer(in)
well-paid [ˌwel 'peɪd]	gut bezahlt
to work out [wɜːk‿'aʊt]	herausfinden, ausarbeiten, lösen

A1

to adapt [ə'dæpt]	bearbeiten
adjective ['ædʒɪktɪv]	Adjektiv
alive [ə'laɪv]	lebend, lebendig, aktiv
degree [dɪ'griː]	akad. Grad (z.B. B.A., M.A.)
enormous [ɪ'nɔːməs]	gewaltig, enorm
to expand [ɪk'spænd]	ausdehnen
fortune ['fɔːtʃuːn]	Glück, Reichtum
to graduate ['grædʒʊeɪt]	Studium beenden, einen akad. Grad erwerben
in the meantime [ɪn ðə 'miːntaɪm]	in der Zwischenzeit
moreover [mɔː'rəʊvə]	überdies, außerdem
natural ['nætʃərəl]	natürlich
noun [naʊn]	Substantiv
risk [rɪsk]	Risiko
sales [seɪlz]	Verkaufszahlen
self-employed [ˌselfɪm'plɔɪd]	selbstständig
to service ['sɜːvɪs]	(Auto, Maschine etc.) warten
to sum up [sʌm 'ʌp]	zusammenfassen
to a great extent [tʊ‿ə ˌgreɪt‿ɪk'stent]	in großem Umfang
turnover ['tɜːnˌəʊvə]	Umsatz
unlimited (company) [ʌnˌlɪmɪtɪd 'kʌmpənɪ]	Gesellschaft mit unbeschränkter Haftung
wage [weɪdʒ]	Lohn

A2

to afford [ə'fɔːd]	sich leisten
to close down [kləʊz 'daʊn]	schließen, zumachen (z.B. Firma)
equipment [ɪ'kwɪpmənt]	Ausrüstung
income ['ɪŋkʌm]	Einkommen

to link [lɪŋk]	verbinden
listener ['lɪsnə]	Zuhörer(in)
main [meɪn]	Haupt-
to make up one's mind [ˌmeɪk‿ʌp wʌnz‿'maɪnd]	sich entschließen
market development [ˌmɑːkɪt dɪ'veləpmənt]	Marktentwicklung
photographer [fə'tɒgrəfə]	Fotograf(in)
to record [rɪ'kɔːd]	aufnehmen (Ton, Video etc.)
to start up (business) [stɑːt ˌʌp 'bɪznɪs]	(Geschäft) anfangen
tense [tens]	Zeitform
thought [θɔːt]	Gedanke
video production ['vɪdɪəʊ prəˌdʌkʃn]	Videoproduktion
wedding ['wedɪŋ]	Hochzeit

C1

amount [ə'maʊnt]	Betrag, Menge
attitude ['ætɪtjuːd]	Haltung, Einstellung
average ['ævərɪdʒ]	Durchschnitt
Bank Holiday [ˌbæŋk 'hɒlɪdeɪ]	(öffentlicher) Feiertag
benefit ['benɪfɪt]	Zuwendung, soziale Leistung
to carry out [ˌkærɪ‿'aʊt]	ausführen
condition [kən'dɪʃn]	Bedingung
Dutch [dʌtʃ]	niederländisch; Niederländer(in)
employee [ɪm'plɔɪiː]	Arbeitnehmer(in), Angestellte(r)
except [ɪk'sept]	außer
to get on with [get‿'ɒn wɪð]	auskommen mit
to have in common [hæv ɪn‿'kɒmən]	gemeinsam haben
to identify [aɪ'dentɪfaɪ]	identifizieren
in general [ɪn 'dʒenərəl]	im Allgemeinen
satisfaction [ˌsætɪs'fækʃn]	Zufriedenheit
satisfied ['sætɪsfaɪd]	zufrieden
sick leave ['sɪk ˌliːv]	krankheitsbedingte Abwesenheit
survey ['sɜːveɪ]	Untersuchung, Umfrage
underpaid [ˌʌndə'peɪd]	unterbezahlt
to vary ['veərɪ]	sich ändern, unterschiedlich sein
working class [ˌwɜːkɪŋ‿'klɑːs]	Arbeiterklasse
working hours ['wɜːkɪŋ‿ˌaʊəz]	Arbeitszeit
workmate ['wɜːkmeɪt]	Arbeitskollege, -kollegin

C2

in order to [ɪn‿'ɔːdə tuː]	um zu
secure [sɪ'kjʊə]	sicher
would like [wʊd 'laɪk]	würde(n) gerne

D

aged ... [eɪdʒd]	... Jahre alt
applicant ['æplɪkənt]	Bewerber(in)
application [ˌæplɪ'keɪʃn]	Bewerbung
to apply [ə'plaɪ]	sich bewerben
assistant [ə'sɪstənt]	Assistent(in)
candidate ['kændɪdət]	Kandidat(in), Bewerber(in)
to care for [keə fɔː]	sich um etw. kümmern
childcare assistant ['tʃaɪldkeər‿əˌsɪstənt]	Kinderpfleger(in)
CV = curriculum vitae [ˌsiː'viː] [kəˌrɪkjələm 'viːtaɪ]	Lebenslauf
data ['deɪtə]	Daten
Dear Madam [dɪə 'mædəm]	Sehr geehrte Dame
Dear Sir [dɪə 'sɜː]	Sehr geehrter Herr
to enclose [ɪn'kləʊz]	beilegen (als Anlage)
essential [ɪ'senʃl]	wesentlich
export assistant ['ekspɔːt‿əˌsɪstənt]	etwa: Außenhandelskaufmann, -frau
foreign ['fɒrən]	ausländisch
leading ['liːdɪŋ]	führend
to look forward to (-ing) [lʊk 'fɔːwəd tuː]	sich freuen auf (Schlussformel im Geschäftsbrief)
nationality [ˌnæʃə'nælətɪ]	Nationalität
overseas [ˌəʊvə'siːz]	in Übersee
reference ['refrəns]	Referenz
to require [rɪ'kwaɪə]	benötigen, erfordern
responsible [rɪ'spɒnsəbl]	verantwortlich
salary ['sælərɪ]	Gehalt
sales department ['seɪls dɪˌpɑːtmənt]	Verkaufsabteilung
similar ['sɪmələ]	ähnlich
step [step]	Schritt, Stufe
task [tɑːsk]	Aufgabe
vacancy ['veɪkənsɪ]	freie Stelle
willing ['wɪlɪŋ]	bereit
with reference to [wɪð 'refrəns tuː]	mit Bezug auf
Yours faithfully [jɔːz‿'feɪθfʊlɪ]	Mit freundlichen Grüßen, Hochachtungsvoll
Yours sincerely [jɔːz‿sɪn'sɪəlɪ]	Mit freundlichen Grüßen

Unit 3

Starter

capital city [ˌkæpɪtl 'sɪtɪ]	Hauptstadt
citizen ['sɪtɪzn]	Bürger(in)
European Commission [ˌjʊərəpɪən kə'mɪʃn]	Europäische Kommission
European Union [ˌjʊərəpɪən‿'juːnjən]	Europäische Union (die EG wurde im November 1993 in EU umbenannt)

A1

bureaucracy [bjʊ'rɒkrəsɪ]	Bürokratie
discount ['dɪskaʊnt]	Rabatt
economics [ˌiːkə'nɒmɪks]	Wirtschaft(swissenschaft)
entrance exam ['entrəns‿ɪgˌzæm]	Aufnahmeprüfung
to fill in [fɪl‿'ɪn]	ausfüllen
to insure [ɪn'ʃʊə]	versichern
to know by name [nəʊ baɪ 'neɪm]	mit Namen kennen
lecture ['lektʃə]	Vorlesung
Poland ['pəʊlənd]	Polen
Polish ['pəʊlɪʃ]	polnisch
to reserve [rɪ'zɜːv]	reservieren
seminar ['semɪnɑː]	Seminar
to take a (language) test [teɪk‿ə 'test]	einen Test machen, an einer (Sprach-) Prüfung teilnehmen
Warsaw ['wɔːsɔː]	Warschau
youth hostel ['juːθ ˌhɒstl]	Jugendherberge

A2

Belgium ['beldʒəm]	Belgien
column ['kɒləm]	Spalte
to come into operation ['kʌm‿ˌɪntʊ‿ˌɒpə'reɪʃn]	in Kraft treten
community [kə'mjuːnətɪ]	Gemeinschaft
currency ['kʌrənsɪ]	Währung
democratic [ˌdemə'krætɪk]	demokratisch
Denmark ['denmɑːk]	Dänemark
election [ɪ'lekʃn]	Wahl
factor ['fæktə]	Faktor
Germany ['dʒɜːmənɪ]	Deutschland
Great Britain [ˌgreɪt‿'brɪtn]	Großbritannien
Greece [griːs]	Griechenland
Ireland ['aɪələnd]	Irland
Italy ['ɪtəlɪ]	Italien
Luxembourg ['lʌksəmbɜːg]	Luxemburg
(the) Netherlands [ðə 'neðələndz]	Niederlande
parliament ['pɑːləmənt]	Parlament
past tense [ˌpɑːst‿'tens]	Vergangenheitsform
peaceful ['piːsfʊl]	friedlich
political [pə'lɪtɪkl]	politisch
process ['prəʊses]	Vorgang, Prozess
single ['sɪŋgl]	einzig, hier: einheitlich
to slow down [sləʊ 'daʊn]	verlangsamen
Spain [speɪn]	Spanien
state [steɪt]	Staat
treaty ['triːtɪ]	Vertrag
unity ['juːnətɪ]	Einheit
wealthy ['welθɪ]	wohlhabend

C1

to afford [ə'fɔːd]	sich leisten
Baltic ['bɔːltɪk]	baltisch; Ostsee
boundary ['baʊndərɪ]	Grenze
communist era ['kɒmjənɪst ˌɪərə]	kommunistisches Zeitalter
cosmopolitan [ˌkɒsmə'pɒlɪtn]	kosmopolitisch
countryside ['kʌntrɪsaɪd]	Land, Landschaft
disappearance [ˌdɪsə'pɪərəns]	Verschwinden
diversity [daɪ'vɜːsətɪ]	Vielfalt, Verschiedenheit
Estonia [es'təʊnɪə]	Estland
glass-walled [ˌglɑːs'wɔːld]	mit Wänden aus Glas
gradually ['grædʒʊəlɪ]	allmählich
high-speed [ˌhaɪ'spiːd]	Hochgeschwindigkeits-
huge [hjuːdʒ]	riesig
individually [ˌɪndɪ'vɪdʒʊəlɪ]	einzeln
leadership ['liːdəʃɪp]	Führung, Leitung
to link; link [lɪŋk]	verbinden; Verbindung
low-cost [ˌləʊ'kɒst]	preiswert
mayor [meə]	Bürgermeister(in)
millennium [mɪ'lenɪəm]	Jahrtausend
Millennium Wheel [mɪ'lenɪəm ˌwiːl]	Millennium-Riesenrad
neglect [nɪ'glekt]	Vernachlässigung
permanent ['pɜːmənənt]	ständig, permanent, dauerhaft
Prague [prɑːg]	Prag
to realise/realize ['rɪəlaɪz]	feststellen, erkennen, bemerken
run-down [ˌrʌn'daʊn]	heruntergekommen
strength [streŋkθ]	Stärke
sunchair ['sʌnˌtʃeə]	Liegestuhl
trading centre ['treɪdɪŋ ˌsentə]	Handelszentrum
to turn into [tɜːn‿'ɪntuː]	verwandeln
view [vjuː]	Blick, Ansicht

C2

above [ə'bʌv]	oben
domestic trip [dəˌmestɪk 'trɪp]	Inlandsausflug
holidaymaker ['hɒlɪdeɪˌmeɪkə]	Urlauber(in)
holidaymaking ['hɒlɪdeɪˌmeɪkɪŋ]	in den Urlaub fahren
least [liːst]	am wenigsten
outbound ['aʊtbaʊnd]	ins Ausland
rate [reɪt]	Quote, Prozentsatz
United Kingdom [juːˌnaɪtɪd 'kɪŋdəm]	Vereinigtes Königreich

D1

conservative [kən'sɜːvətɪv]	konservativ
constituency [kən'stɪtjʊənsɪ]	Wahlkreis
democrat ['deməkræt]	Demokrat(in)
former ['fɔːmə]	ehemalig
labour ['leɪbə]	Arbeit
voter ['vəʊtə]	Wähler(in)
voting system ['vəʊtɪŋ ˌsɪstəm]	Wahlsystem

D2

to appoint [ə'pɔɪnt]	ernennen, bestimmen
cabinet ['kæbɪnət]	Kabinett
to delay [dɪ'leɪ]	verschieben, aufschieben
diagram ['daɪəgræm]	Diagramm
to elect [ɪ'lekt]	wählen
hereditary peers [hɪˌredɪtərɪ 'pɪəz]	Peer mit vererbtem Adelstitel
House of Commons [ˌhaʊs‿əv 'kɒmənz]	Unterhaus
House of Lords [ˌhaʊs‿əv 'lɔːdz]	Oberhaus
Law Lord ['lɔː ˌlɔːd]	Mitglied des Oberhauses mit besonderer Verantwortung in Rechtsfragen
legislation [ˌledʒɪs'leɪʃn]	Gesetzgebung
life peers [ˌlaɪf 'pɪəz]	Peer auf Lebenszeit
on recommendation [ɒn ˌrekəmen'deɪʃn]	auf Empfehlung
previous ['priːvɪəs]	vorhergehend, früher
prime minister [ˌpraɪm‿'mɪnɪstə]	Premierminister(in)
sovereign ['sɒvrɪn]	Souverän, Herrscher(in)
to vote; vote [vəʊt]	wählen; Stimme

Unit 4

Starter

communications [kəˌmjuːnɪ'keɪʃnz]	Telekommunikation, Kommunikationswege
device [dɪ'vaɪs]	Gerät

A1

actual ['æktʃʊəl]	tatsächlich
apparatus [ˌæpə'reɪtəs]	Gerät
appliance [ə'plaɪəns]	Gerät
to call up [kɔːl‿'ʌp]	anrufen, aufrufen
to combine [kəm'baɪn]	verbinden, vereinigen
complex ['kɒmpleks]	komplex, kompliziert
complicated ['kɒmplɪkeɪtɪd]	kompliziert
consumer [kən'sjuːmə]	Verbraucher(in)
to crash [kræʃ]	zusammenbrechen
data ['deɪtə]	Daten
digital ['dɪdʒɪtl]	digital
document ['dɒkjʊmənt]	Dokument, Unterlage
easy-to-use [ˌiːzɪ tʊ‿'juːz]	einfach zu bedienen
electronic [ˌɪlek'trɒnɪk]	elektronisch
multi- ['mʌltɪ]	Mehr-, Viel-
network ['netwɜːk]	Netzwerk
paradise ['pærədaɪs]	Paradies
poverty ['pɒvətɪ]	Armut
screen [skriːn]	Bildschirm, Leinwand
to store [stɔː]	speichern
transmission [trænz'mɪʃn]	Übertragung
to transmit [trænz'mɪt]	senden
TV set [ˌtiː'viː ˌset]	Fernsehgerät
user ['juːzə]	Benutzer(in)
user-friendly [ˌjuːzə'frendlɪ]	benutzerfreundlich
virtual ['vɜːtʃʊəl]	virtuell
wireless ['waɪələs]	drahtlos

A2

access ['ækses]	Zugang
bar graph ['bɑːgrɑːf]	Säulendiagramm
broadband ['brɔːdbænd]	Breitband
figure ['fɪgə]	Zahl, Figur
graph [grɑːf]	Diagramm, Schaubild
household ['haʊshəʊld]	Haushalt
mobile phone [ˌməʊbaɪl 'fəʊn]	Handy
percentage [pə'sentɪdʒ]	prozentualer Anteil

C

according to [ə'kɔːdɪŋ‿tuː]	(seiner/ihrer) Meinung nach, laut
advertiser ['ædvəˌtaɪzə]	Inserent(in), Auftraggeber(in) von Werbesendungen
airtime ['eətaɪm]	Sendezeit
background ['bækgraʊnd]	Hintergrund
best known [ˌbest‿'nəʊn]	bekannteste(r, s)
campaign [kæm'peɪn]	Kampagne, Werbefeldzug
commercial [kə'mɜːʃl]	Werbespot
to decrease [dɪ'kriːs]	abnehmen, vermindern
expenditure [ɪk'spendɪtʃə]	Ausgabe(n), Aufwand
to fail [feɪl]	scheitern
habit ['hæbɪt]	Gewohnheit
logo ['ləʊgəʊ]	Logo, (Firmen-)Emblem
on the one hand [ɒn‿ðə 'wʌn‿ˌhænd]	auf der einen Seite
outdoor [ˌaʊt'dɔː]	draußen, im Freien
to overestimate [ˌəʊvər'estɪmeɪt]	überschätzen
per head [pə 'hed]	pro Kopf
pleased [pliːzd]	zufrieden, erfreut
to recognize ['rekəgnaɪz]	(wieder)erkennen
remote control [rɪˌməʊt‿kən'trəʊl]	Fernbedienung
share [ʃeə]	Anteil

to switch [swɪtʃ]	umschalten, schalten
tennis serve ['tenɪs ˌsɜːv]	Aufschlag (Tennis)
value ['væljuː]	Wert
viewer ['vjʊə]	Zuschauer(in)
viewpoint ['vjuːpɔɪnt]	Standpunkt
whereas [weə'ræz]	während
worldwide [ˌwɜːld'waɪd]	weltweit

D1

polite [pə'laɪt]	höflich
superb [sjuː'pɜːb]	erstklassig
superfast [ˌsjuːpə'fɑːst]	sehr schnell

D2

advert ['ædvɜːt]	Kurzform von "advertisement" = Werbeanzeige
to aim (at) [eɪm]	zielen (auf)
to associate [ə'səʊʃɪeɪt]	assoziieren, in Verbindung bringen
brand name ['brænd‿ˌneɪm]	Markenname
lifestyle ['laɪfstaɪl]	Lebensstil
to refer to [rɪ'fɜː tuː]	sich beziehen auf
slogan ['sləʊgn]	Slogan, Werbespruch
target group ['tɑːgɪt‿ˌgruːp]	Zielgruppe

Unit 5

Starter

to reduce [rɪdjuːs]	verringern, senken
view [vjuː]	Sicht, Ansicht, Aussicht

A1

congested [kən'dʒestɪd]	überfüllt, verstopft
congestion [kən'dʒestʃn]	Stau
day by day [ˌdeɪ baɪ 'deɪ]	Tag für Tag
fumes [fjuːmz]	Rauch, Dämpfe, Abgase
to get used to [get 'juːst‿tuː]	sich gewöhnen an
innovation [ˌɪnə'veɪʃn]	Neuerung
to look into [lʊk‿'ɪntuː]	untersuchen, prüfen
major ['meɪdʒə]	bedeutend
motorway ['məʊtəweɪ]	Autobahn
navigation system [ˌnævɪ'geɪʃn ˌsɪstəm]	Navigationssystem
overcrowded [ˌəʊvə'kraʊdɪd]	überfüllt
rush-hour ['rʌʃˌaʊə]	Hauptverkehrszeit
technological [ˌteknə'lɒdʒɪkl]	technologisch, technisch
traffic jam ['træfɪk ˌdʒæm]	Verkehrsstau
tram [træm]	Straßenbahn

A2

accent ['æksənt]	Akzent
additional [ə'dɪʃənl]	zusätzlich
aid [eɪd]	Hilfe, Hilfsmittel
audio ['ɔːdɪəʊ]	audio, Hör-
available [ə'veɪləbl]	verfügbar, vorhanden
to avoid [ə'vɔɪd]	vermeiden
button ['bʌtn]	Knopf
command [kə'mɑːnd]	Befehl
destination [ˌdestɪ'neɪʃn]	Reiseziel, Zielort
display [dɪ'spleɪ]	Anzeige
diversion [daɪ'vɜːʃn]	Umleitung
due to [djuː tuː]	aufgrund von
to enter ['entə]	eingeben (Daten)
to equip [ɪ'kwɪp]	ausrüsten
feature ['fiːtʃə]	Merkmal, Bestandteil
fog [fɒg]	Nebel
glance [glɑːns]	Blick
GPS (Global Positioning System) [ˌdʒiːpiː'es] [ˌgləʊbl pə'zɪʃənɪŋ ˌsɪstəm]	GPS-Navigationssystem
to guide [gaɪd]	leiten, führen
to install [ɪn'stɔːl]	einbauen
interval ['ɪntəvl]	Abstand
location [ləʊ'keɪʃn]	Lage, Standort
to monitor ['mɒnɪtə]	überwachen
network ['netwɜːk]	Netz
to obtain [ɒb'teɪn]	bekommen, erhalten
petrol station ['petrl ˌsteɪʃn]	Tankstelle
to pinpoint ['pɪnpɔɪnt]	genau anzeigen, genau bestimmen
to programme ['prəʊgræm]	programmieren
to recalculate [ˌriː'kælkjəleɪt]	neu berechnen
to reduce [rɪ'dʒuːs]	verringern, senken
to regulate ['regjəleɪt]	regulieren
to rely on [rɪ'laɪ‿ɒn]	sich verlassen auf
to rewrite [ˌriː'raɪt]	neu schreiben, umschreiben
road works ['rəʊdˌwɜːks]	Straßenbauarbeiten
to run out of sth. [rʌn‿'aʊt‿əv ˌsʌmθɪŋ]	ausgehen, etwas nicht mehr haben
satellite ['sætəlaɪt]	Satellit
sensor ['sensə]	Sensor, Fühler
slow moving traffic ['sləʊˌmuːvɪŋ 'træfɪk]	zäh fließender Verkehr
steering wheel ['stɪərɪŋ‿ˌwiːl]	Lenkrad
suburban [sə'bɜːbn]	Vorort-
to suggest [sə'dʒest]	vorschlagen
to take care [teɪk‿'keə]	aufpassen
touch [tʌtʃ]	Berührung
traffic flow ['træfɪk ˌfləʊ]	Verkehrsfluss
traffic jam ['træfɪk ˌdʒæm]	Verkehrsstau
to transmit [trænz'mɪt]	übermitteln

user ['ju:zə]	Benutzer(in)
voice recognition ['vɔɪs ˌrekəg'nɪʃn]	Spracherkennung
to worry ['wʌrɪ]	sich sorgen, sich beunruhigen

C1

attractive [ə'træktɪv]	reizvoll, anziehend, attraktiv
commuter [kə'mju:tə]	Pendler(in)
convenient [kən'vi:nɪənt]	günstig, praktisch
crazy ['kreɪzɪ]	verrückt
ecological [ˌi:kə'lɒdʒɪkl]	ökologisch, umweltfreundlich
electrically-driven [ɪˌlektrɪklɪ 'drɪvən]	elektrisch angetrieben
environmentalist [ɪnˌvaɪrən'mentəlɪst]	Umweltschützer(in)
to estimate ['estɪmeɪt]	schätzen
expert ['ekspɜ:t]	Experte, Expertin
to face sth. ['feɪs‿ˌsʌmθɪŋ]	mit etw. konfrontiert sein
fixed-route [ˌfɪkst'ru:t]	schienengebunden
flexible ['fleksɪbl]	flexibel
freedom ['fri:dəm]	Freiheit
to get rid of [get 'rɪd‿əv]	loswerden, abschaffen
heading ['hedɪŋ]	Überschrift
individual [ˌɪndɪ'vɪdʒʊəl]	individuell
in favour of [ɪn 'feɪvər‿əv]	zugunsten von
inner ['ɪnə]	Innen-, inner-
to make sense [meɪk 'sens]	einen Sinn ergeben
medium-sized [ˌmi:djəm 'saɪzd]	mittelgroß
peak time ['pi:k ˌtaɪm]	Hauptverkehrszeit
planner ['plænə]	Planer(in)
progress ['prəʊgres]	Fortschritt
to provide [prə'vaɪd]	sorgen für, liefern, bereitstellen
to put into practice [ˌpʌt‿ɪntʊ 'præktɪs]	in die Praxis umsetzen, realisieren
rail [reɪl]	Schiene
sort [sɔ:t]	Art, Sorte
thanks to ['θæŋks tu:]	dank
tramway ['træmweɪ]	Straßenbahngleis
underground ['ʌndəgraʊnd]	Untergrund, U-Bahn
unpopular [ˌʌn'pɒpjʊlə]	unpopulär

C2

definite ['defɪnɪt]	eindeutig, bestimmt
to suffer ['sʌfə]	leiden

D1

motorist ['məʊtərɪst]	Autofahrer(in)

D2

base [beɪs]	Basis, Stützpunkt, Ausgangspunkt
by rail [baɪ 'reɪl]	mit der Bahn
coaching days ['kəʊtʃɪŋ ˌdeɪz]	Postkutschenzeit
to concentrate on ['kɒnsəntreɪt‿ɒn]	konzentrieren auf
continental [ˌkɒntɪ'nentl]	kontinental
crossroads ['krɒsrəʊdz]	Kreuzung
intercontinental [ˌɪntəkɒntɪ'nentl]	interkontinental
on top of [ɒn 'tɒp‿əv]	zusätzlich zu
to operate ['ɒpəreɪt]	verkehren, bedienen
promotion [prə'məʊʃn]	Werbung, Beförderung
town hall [ˌtaʊn 'hɔ:l]	Rathaus
via ['vaɪə]	über

Unit 6

Starter

chart [tʃɑ:t]	Tabelle, Schaubild
composting ['kɒmpɒstɪŋ]	Kompostierung
energy recovery ['enədʒɪ rɪˌkʌvərɪ]	Energierückgewinnung
landfill ['lændfɪl]	Mülldeponie
plastic ['plæstɪk]	Kunststoff, Plastik
recovery [rɪ'kʌvərɪ]	Wiedergewinnung
ton [tʌn]	Tonne
waste [weɪst]	Müll, Abfall

A1

acid rain [ˌæsɪd‿'reɪn]	saurer Regen
advertising campaign ['ædvətaɪzɪŋ‿kæmˌpeɪn]	Werbekampagne
at all costs [ət‿ˌɔ:l 'kɒsts]	um jeden Preis
bulb [bʌlb]	(Glüh)birne
conscious ['kɒnʃəs]	bewusst
consumer [kən'sju:mə]	Verbraucher(in)
consumption [kən'sʌmpʃn]	Verbrauch
to contribute [kən'trɪbju:t]	beitragen
contribution [ˌkɒntrɪ'bju:ʃn]	Beitrag
disposal [dɪs'pəʊsl]	Beseitigung, Entsorgung
emission [ɪ'mɪʃn]	Emission, Ausstoß
emphasis ['emfəsɪs]	Betonung
to emphasize ['emfəsaɪz]	betonen
environmental [ɪnˌvaɪrən'mentl]	Umwelt-
environmentalist [ɪnˌvaɪrən'mentəlɪst]	Umweltschützer(in)
in conclusion [ɪn‿kən'klu:ʒn]	abschließend
to influence ['ɪnflʊəns]	beeinflussen
to insulate ['ɪnsjəleɪt]	isolieren

insulation [ˌɪnsjəˈleɪʃn]	Isolierung
lighting [ˈlaɪtɪŋ]	Beleuchtung
to manufacture [ˌmænjəˈfæktʃə]	herstellen
manufacturer [ˌmænjəˈfæktʃərə]	Hersteller(in)
to meet the growing demand [ˌmiːt‿ðə ˌɡrəʊɪŋ‿dɪˈmɑːnd]	der wachsenden Nachfrage gerecht werden
non-returnable [ˌnɒnrɪˈtɜːnəbl]	Einweg-
once again [wʌns‿əˈɡen]	noch einmal
ozone hole [ˈəʊzəʊn ˌhəʊl]	Ozonloch
packaging [ˈpækɪdʒɪŋ]	Verpackung
power station [ˈpaʊə ˌsteɪʃn]	Kraftwerk, Elektrizitätswerk
range [reɪndʒ]	Sortiment, Umfang, Reichweite
reaction [rɪˈækʃn]	Reaktion
recyclable [ˌriːˈsaɪkləbl]	wiederverwertbar
reduction [rɪˈdʌkʃn]	Verminderung
to refuse [rɪˈfjuːz]	ablehnen, sich weigern
responsibility [rɪˌspɒnsəˈbɪlətɪ]	Verantwortung
retailer [ˈriːteɪlə]	Einzelhändler(in)
returnable [rɪˈtɜːnəbl]	Mehrweg-
rubbish [ˈrʌbɪʃ]	Abfall, Müll
so-called [ˌsəʊˈkɔːld]	so genannt
to turn to [ˈtɜːn tuː]	sich zuwenden
wasteful [ˈweɪstfʊl]	verschwenderisch

A2

additional [əˈdɪʃənl]	zusätzlich
all over the world [ɔːl‿ˌəʊvə ðə ˈwɜːld]	in der ganzen Welt
CFC [ˌsiːefˈsiː]	FCKW (Fluorchlorkohlenwasserstoff)
chemist [ˈkemɪst]	Chemiker(in)
convinced [kənˈvɪnst]	überzeugt
direct speech [daɪˌrekt ˈspiːtʃ]	direkte Rede
disagreement [ˌdɪsəˈɡriːmənt]	abweichende Meinung, Meinungsverschiedenheit
elsewhere [ˌelsˈweə]	woanders
fuel [fjʊəl]	Brennstoff, Kraftstoff
furthermore [ˈfɜːðəmɔː]	außerdem, ferner
improvement [ɪmˈpruːvmənt]	Verbesserung
to maintain [meɪnˈteɪn]	behaupten
moreover [mɔːˈrəʊvə]	überdies, zudem, ferner
ought to [ˈɔːt‿tuː]	sollte(n)
to take the view [teɪk‿ðə ˈvjuː]	der Ansicht sein

C

(not) any more [nɒt / ˌenɪ‿ˈmɔː]	nicht länger
approach [əˈprəʊtʃ]	Ansatz, Annäherung
to attack [əˈtæk]	in Angriff nehmen, angreifen
card [kɑːd]	Karte
to code; code [kəʊd]	kodieren; Kode
to come true [kʌm ˈtruː]	wahr werden
dealer [ˈdiːlə]	Händler(in)
to dismantle [dɪsˈmæntl]	zerlegen, demontieren
to dump; dump [dʌmp]	(Müll) abladen; Müllkippe
Friends of the Earth [ˌfrendz‿əv‿ðɪ‿ˈɜːθ]	Name einer Umweltschutzorganisation
to get rid of [ɡet ˈrɪd‿əv]	loswerden
left [left]	übrig
to melt [melt]	schmelzen
over and over again [ˌəʊvər‿ən‿ˌəʊvər‿əˈɡen]	immer wieder
paraphrase [ˈpærəfreɪz]	Umschreibung
precise [prɪˈsaɪs]	genau
raw [rɔː]	Roh-, roh
to re-use [ˌriːˈjuːz]	wiederverwenden
to recycle [ˌriːˈsaɪkl]	wiederverwerten
scrap [skræp]	Schrott, Fetzen
to separate [ˈsepəreɪt]	trennen
to shift [ʃɪft]	verlagern
to substitute; substitute [ˈsʌbstɪtjuːt]	ersetzen; Ersatz
supplier [səˈplaɪə]	Lieferant(in)
to take to pieces [ˌteɪk tʊ ˈpiːsɪz]	zerlegen
unknown [ˌʌnˈnəʊn]	unbekannt
unlikely [ʌnˈlaɪklɪ]	unwahrscheinlich
waste management [ˈweɪst ˌmænɪdʒmənt]	Abfallentsorgung

D1

countryside [ˈkʌntrɪsaɪd]	Landschaft
to create [kriːˈeɪt]	schaffen, verursachen
deposit [dɪˈpɒzɪt]	Lagerstätte, Guthaben
ecology [ɪˈkɒlədʒɪ]	Ökologie
fear [fɪə]	Furcht
mining [ˈmaɪnɪŋ]	Bergbau
nature [ˈneɪtʃə]	Natur
ore [ɔː]	Erz
recording [rɪˈkɔːdɪŋ]	Aufnahme
summary [ˈsʌmərɪ]	Zusammenfassung
technologist [tekˈnɒlədʒɪst]	Techniker(in)
zinc [zɪŋk]	Zink

D2

attraction [əˈtrækʃn]	Anziehungskraft
coastal [ˈkəʊstl]	Küsten-
concerned [kənˈsɜːnd]	betroffen, interessiert
housing estate [ˈhaʊzɪŋ‿ɪˌsteɪt]	Siedlung
issue [ˈɪʃuː]	Thema, Frage
major road [ˌmeɪdʒə‿ˈrəʊd]	Hauptstraße

MP (Member of Parliament) [ˌem'piː] [ˌmembər‿əv 'pɑːləmənt]	Mitglied des Parlaments
primary school ['praɪmərɪ ˌskuːl]	Grundschule
public library [ˌpʌblɪk 'laɪbrərɪ]	Stadtbücherei
reported speech [rɪˌpɔːtɪd 'spiːtʃ]	indirekte Rede
role [rəʊl]	Rolle
seaside resort ['siːsaɪd rɪˌzɔːt]	Seebad
surface ['sɜːfɪs]	Oberfläche
tax [tæks]	Steuer
unemployed [ˌʌnɪm'plɔɪd]	arbeitslos

Unit 7

Starter

fast running [ˌfɑːst 'rʌnɪŋ]	schnell fließend
machinery [mə'ʃiːnərɪ]	Maschinen
peat bog ['piːt ˌbɒg]	Torfmoor
resources [rɪ'sɔːsɪz] [rɪ'zɔːsɪz]	Boden-, Naturschätze, Mittel

A1

antonym ['æntənɪm]	Antonym, Wort mit gegensätzlicher Bedeutung
availability [əˌveɪlə'bɪlətɪ]	Vorhandensein, Verfügbarkeit
billion ['bɪljən]	Billion, (AE: Milliarde)
citizen ['sɪtɪzn]	Bürger(in)
developing country [dɪˌveləpɪŋ‿'kʌntrɪ]	Entwicklungsland
fossil fuels [ˌfɒsl 'fjʊəlz]	fossile Brennstoffe
geographical [ˌdʒɪə'græfɪkl]	geografisch
global warming [ˌgləʊbl 'wɔːmɪŋ]	weltweiter Temperaturanstieg
to harm [hɑːm]	schädigen
impression [ɪm'preʃn]	Eindruck
industrialized [ɪn'dʌstrɪəlaɪzd]	industrialisiert
mankind [mæn'kaɪnd]	Menschheit
plentiful ['plentɪfʊl]	reichlich, häufig
plenty ['plentɪ]	viel, eine Menge
to predict [prɪ'dɪkt]	vorhersagen
secure [sɪ'kjʊə]	sicher
significant [sɪg'nɪfɪkənt]	bedeutend, bedeutsam
slight [slaɪt]	leicht, schwach, gering
solar power [ˌsəʊlə 'paʊə]	Sonnenenergie
source [sɔːs]	Quelle
standard of living [ˌstændəd‿əv 'lɪvɪŋ]	Lebensstandard
to take action [teɪk 'ækʃn]	etwas unternehmen
total ['təʊtl]	gesamt
to transfer [træns'fɜː]	übertragen
trillion ['trɪljən]	Trillion (AE: Billion)
wealth [welθ]	Reichtum
what ... like? [wɒt ... 'laɪk]	wie (ist er/sie/es) ...?

A2

BTU British Thermal Unit [ˌbiːtiː'juː] [ˌbrɪtɪʃ ˌθɜːml 'juːnɪt]	BTU = brit. Wärmeeinheit
constant ['kɒnstənt]	ständig, gleichbleibend
to drop [drɒp]	fallen, fallen lassen
gradual ['grædʒʊəl]	allmählich, sanft (ansteigen/abfallen)
graph [grɑːf]	Schaubild, mathematische Kurve
projection [prə'dʒekʃn]	Prognose, Projektion
sharp [ʃɑːp]	scharf, deutlich, steil
stable ['steɪbl]	stabil, dauerhaft
steady ['stedɪ]	stabil, gleichbleibend
substantial [səb'stænʃl]	beträchtlich, erheblich
temperature ['temprətʃə]	Temperatur

C

to argue ['ɑːgjuː]	behaupten
as a result [əz‿ə rɪ'zʌlt]	folglich
beer [bɪə]	Bier
cable ['keɪbl]	Kabel
cell [sel]	Zelle
consequently ['kɒnsɪkwəntlɪ]	folglich
to dig up [dɪg‿'ʌp]	(aus)graben
domestic [də'mestɪk]	häuslich
e.g. [ˌiː'dʒiː] [fər‿ɪg'zɑːmpl]	z. B. (zum Beispiel)
fossil fuels [ˌfɒsl 'fjʊəlz]	fossile Brennstoffe
fuel cell ['fjʊəl ˌsel]	Brennstoffzelle
to generate ['dʒenəreɪt]	erzeugen
grid [grɪd]	Versorgungsnetz
highway ['haɪweɪ]	Landstraße
hydro ['haɪdrəʊ]	Wasser-
hydrogen-based [ˌhaɪdrədʒn'beɪst]	auf Wasserstoffbasis arbeitend
in the mid [ɪn‿ðə 'mɪd]	in der Mitte von, Mitte der ...
introductory [ˌɪntrə'dʌktərɪ]	einleitend
to join [dʒɔɪn]	sich anschließen
key point ['kiː ˌpɔɪnt]	springender Punkt
to light (lit, lit) [laɪt lɪt lɪt]	anzünden, (be)leuchten
maintenance ['meɪntənəns]	Wartung, Instandhaltung
maker ['meɪkə]	Hersteller(in)
off-the-grid [ˌɒf‿ðə 'grɪd]	nicht an das Versorgungsnetz angeschlossen
panel ['pænl]	Schalttafel
photovoltaic (pv) [ˌfəʊtəʊvɒl'teɪɪk] [ˌpiː'viː]	fotoelektrisch

powered ['paʊəd] angetrieben, betrieben
power line ['paʊə laɪn] (Stark-)Stromleitung
power plant ['paʊə ˌplɑːnt] Kraftwerk
power station ['paʊə ˌsteɪʃn] Kraftwerk
renewable energy [rɪˌnjuːəbl 'enədʒɪ] erneuerbare Energie
research ['riːsɜːtʃ] Forschung
scientist ['saɪəntɪst] Wissenschaftler(in)
shower ['ʃaʊə] Dusche
to state [steɪt] darlegen, vortragen, nennen
subject matter ['sʌbdʒɪkt ˌmætə] Stoff, Inhalt
to supply [sə'plaɪ] versorgen
traditional [trə'dɪʃənl] traditionell, herkömmlich
yearly ['jɪəlɪ] jährlich

D1

agricultural [ˌægrɪ'kʌltʃərəl] landwirtschaftlich
biomass ['baɪəʊmæs] Biomasse
conversation [ˌkɒnvə'seɪʃn] Unterhaltung, Gespräch
conveyor [kən'veɪə] Förderband
delivery hall [dɪ'lɪvrɪ ˌhɔːl] Anlieferungshalle
fertilizer ['fɜːtəlaɪzə] Dünger
to fix [fɪks] vorhaben, organisieren
furnace ['fɜːnɪs] Ofen
generator ['dʒenəreɪtə] Generator, Lichtmaschine
litter ['lɪtə] Streu, Stroh
manure [mə'njʊə] Dung
megawatt ['megəwɒt] Megawatt
neat [niːt] ordentlich, gelungen, schlau
output ['aʊtpʊt] (Produktions-)Leistung
pit [pɪt] Grube
poultry ['pəʊltrɪ] Geflügel
steam [stiːm] Dampf
storage ['stɔːrɪdʒ] Lagerung, Aufbewahrung
straight in [streɪt‿'ɪn] direkt hinein
to tip [tɪp] kippen
turbine ['tɜːbaɪn] Turbine

D2

ash [æʃ] Asche
conveyor belt [kən'veɪə ˌbelt] Förderband, Fließband
employee [ɪm'plɔɪiː] Angestellte(r)
greenhouse ['griːnhaʊs] Treibhaus, Gewächshaus
manpower ['mænˌpaʊə] Arbeitskraft

Unit 8

A1

again and again [əˌgen‿ən‿ə'gen] immer wieder
all kind(s) of [ɔːl 'kaɪndz‿əv] alle Arten von, alle möglichen
to automate ['ɔːtəmeɪt] automatisieren
automatic [ˌɔːtə'mætɪk] automatisch
available [ə'veɪləbl] verfügbar, vorhanden
CAD [ˌsiːeɪ'diː] (Computer Aided Design) [kəm'pjuːtər‿ˌeɪdɪd‿dɪ'zaɪn] Computerunterstützte Zeichnung
CAM [ˌsiː eɪ'em] (Computer Aided Manufacturing) [kəm'pjuːtər‿ˌeɪdɪd mænjə'fæktʃərɪŋ] Computerunterstützte Fertigung
civilized ['sɪvɪlaɪzd] zivilisiert
CIM [ˌsiː aɪ'em] (Computer Integrated Manufacturing) [kəm'pjuːtər‿ˌɪntɪgreɪtɪd mænjə'fæktʃərɪŋ] Computerintegrierte Fertigung
designer [dɪ'zaɪnə] Zeichner(in), Designer(in)
economical [ˌiːkə'nɒmɪkl] sparsam, wirtschaftlich
to equip [ɪ'kwɪp] ausrüsten
etc. (et cetera) [ˌiː tiː'siː] [ɪt'setərə] und so weiter
fault [fɔːlt] Fehler, Schuld, Mangel
feedback ['fiːdbæk] Rückmeldung, Feedback
latest ['leɪtɪst] der, die, das Neueste
to link together [lɪŋk‿tə'geðə] miteinander verbinden
particular [pə'tɪkjələ] besondere(r,s)
productivity [ˌprɒdʌk'tɪvətɪ] Produktivität, Leistungsfähigkeit
research department ['riːsɜːtʃ dɪˌpɑːtmənt] Forschungsabteilung
robot ['rəʊbɒt] Roboter, vollautomatische Vorrichtung
sensor ['sensə] Fühler, Sensor
shop floor [ˌʃɒp‿'flɔː] Produktionsstätte, -halle
to simulate ['sɪmjəleɪt] simulieren
term [tɜːm] Ausdruck, Wort
to touch; touch [tʌtʃ] berühren; Berührung
unsuitable [ʌn'sjuːtəbl] unpassend, ungeeignet

A2

to assemble [ə'sembl] zusammenbauen
bumper ['bʌmpə] Stoßstange
car body ['kɑː ˌbɒdɪ] Karosserie
dashboard ['dæʃˌbɔːd] Armaturenbrett
to dip [dɪp] (in Farbe) eintauchen
engine ['ɛndʒɪn] Motor
exhaust pipe [ɪg'zɔːst‿ˌpaɪp] Auspuffrohr

to fit [fɪt]	einbauen, anbringen
headlight ['hedlaɪt]	Scheinwerfer
to spray [spreɪ]	sprühen, spritzen
to weld [weld]	schweißen
windscreen ['wɪndˌskriːn]	Windschutzscheibe

C1

to arrange [ə'reɪndʒ]	aufstellen, planen, veranlassen
arrangement [ə'reɪndʒmənt]	Vereinbarung, Arrangement
customer-orientated [ˌkʌstəmər‿'ɔːrɪənteɪtɪd]	kundenorientiert
decision-making process [dɪ'sɪʒn‿ˌmeɪkɪŋ ˌprəʊses]	Entscheidungsprozess
to dominate ['dɒmɪneɪt]	beherrschen, dominieren
flexibility [ˌfleksə'bɪlətɪ]	Flexibilität
inflexible [ɪn'fleksəbl]	unflexibel
Japanese [ˌdʒæpə'niːz]	japanisch; Japaner(in)
lean management [ˌliːn 'mænɪdʒmənt]	„schlankes" Management
monotonous [mə'nɒtənəs]	eintönig, monoton
opportunity [ɒpə'tjuːnətɪ]	Gelegenheit, Möglichkeit
to perform [pə'fɔːm]	ausführen, vollbringen
personnel [ˌpɜːsə'nel]	Personal
repetitive [rɪ'petətɪv]	(sich) wiederholend, eintönig
to replace [rɪ'pleɪs]	ersetzen
schedule ['ʃedjuːl]	(Zeit-, Fahr-)Plan, Programm
style [staɪl]	Stil, Art
to suit [sjuːt]	passen
task [tɑːsk]	Aufgabe
team work ['tiːmwɜːk]	Teamarbeit
union ['juːnjən]	Gewerkschaft
unit ['juːnɪt]	Einheit, Gruppe

C2

DVD [ˌdiːviː'diː] (recorder)	DVD-Rekorder (-Aufnahmegerät)
infinitive [ɪn'fɪnətɪv]	Infinitiv, Grundform
to insert [ɪn'sɜːt]	einlegen (Diskette, Kassette)
modal ['məʊdl]	Hilfsverb
model ['mɒdl]	Modell
passive ['pæsɪv]	Passiv
to press [pres]	drücken
to record [rɪ'kɔːd]	aufnehmen
recording instruction [rɪ'kɔːdɪŋ‿ɪnˌstrʌkʃn]	Aufnahmeanleitung
to select [sə'lekt]	auswählen
to switch on [swɪtʃ‿'ɒn]	anstellen

D1

handbook ['hændbʊk]	Handbuch
installation [ˌɪnstə'leɪʃn]	Installation, Anschluss
to print [prɪnt]	drucken
printer ['prɪntə]	Drucker
resolution [ˌrezə'luːʃn]	Auflösung

D2

cartridge ['kɑːtrɪdʒ]	Kassette, Patrone
connector [kə'nektə]	Verbindungselement, Stecker
cover ['kʌvə]	Abdeckung
ink [ɪŋk]	Tinte
ink jet printer ['ɪŋkdʒet ˌprɪntə]	Tintenstrahldrucker
interface ['ɪntəfeɪs]	Schnittstelle
to make sure [meɪk‿'ʃʊə]	sich überzeugen
to mix up [mɪks‿'ʌp]	vertauschen
to remove [rɪ'muːv]	entfernen
to screw; screw [skruː]	schrauben; Schraube
to snap [snæp]	einrasten
to take off [teɪk‿'ɒf]	abnehmen
tape [teɪp]	Band, Klebestreifen
to tighten ['taɪtn]	(fest) anziehen, verschärfen

Unit 9

Starter

financial services [faɪˌnənʃl 'sɜːvɪsɪz]	Finanzdienstleistungen
legal services [ˌliːgl 'sɜːvɪsɪz]	Rechtsberatungen, -dienstleistungen
printing ['prɪntɪŋ]	Druckwesen
publishing ['pʌblɪʃɪŋ]	Verlagswesen

A1

shipbuilding ['ʃɪpˌbɪldɪŋ]	Schiffbau
shipyard ['ʃɪpjɑːd]	Werft
steel [stiːl]	Stahl

A2

to achieve [ə'tʃiːv]	erreichen, erzielen
arms [ɑːmz]	Waffen
to be located [biː ləʊ'keɪtɪd]	sich befinden, liegen
broadband ['brɔːdbænd]	Breitband
corporation [ˌkɔːpə'reɪʃn]	Gesellschaft, Körperschaft
coverage ['kʌvərɪdʒ]	hier: Ausstattung
data processing ['deɪtə ˌprəʊsesɪŋ]	Datenverarbeitung
decade ['dekeɪd]	Jahrzehnt, Dekade
entrepreneurship [ˌɒntrəprə'nɜːʃɪp]	Unternehmertum, unternehmerische Vorhaben

former ['fɔːmə]	früher
innovation [ˌɪnə'veɪʃn]	Innovation, Neuerung
large-scale [ˌlɑːdʒ'skeɪl]	groß angelegt, in großem Umfang
to let [let]	vermieten
light industry [ˌlaɪt 'ɪndʌstrɪ]	Leichtindustrie
managing director [ˌmænɪdʒɪŋ‿daɪ'rektə]	Geschäftsführer(in)
to occupy ['ɒkjəpaɪ]	einnehmen, besetzen, beschäftigen
official [ə'fɪʃl]	offiziell; Beamte(r), Funktionär(in)
outstanding [ˌaʊt'stændɪŋ]	hervorragend, außergewöhnlich
to provide [prə'vaɪd]	bereitstellen
range [reɪndʒ]	Reihe, Auswahl
scheme [skiːm]	Plan, Programm, Projekt
site [saɪt]	Grundstück, Sitz, Standort
square [skweə]	Quadrat
to take over [teɪk‿'əʊvə]	übernehmen
wired ['waɪəd]	verkabelt

C

to account for [ə'kaʊnt fɔː]	beitragen
to add [æd]	hinzufügen
to announce [ə'naʊns]	bekannt geben
annual output [ˌænjʊəl‿'aʊtpʊt]	Jahresproduktion
appeal [ə'piːl]	Anziehungskraft
assembly line [ə'semblɪ ˌlaɪn]	Fließband
average increase [ˌævərɪdʒ‿'ɪnkriːs]	durchschnittlicher Anstieg
bagless ['bægləs]	hier: ohne Staubsaugerbeutel
by contrast [baɪ 'kɒntrɑːst]	im Gegensatz
call centre ['kɔːl ˌsentə]	Callcenter, telefon. Auskunftsdienst
to contrast [kən'trɑːst]	vergleichen
desperately ['despərətlɪ]	verzweifelt
economical [ˌiːkə'nɒmɪkl]	wirtschaftlich, sparsam
to eliminate [ɪ'lɪmɪneɪt]	beseitigen, entfernen
estimated value [ˌestɪmeɪtɪd 'væljuː]	geschätzter Wert
icon ['aɪkən]	Idol, Ikone
innovative ['ɪnəvətɪv] [ɪ'nɒvətɪv]	innovativ
inventor [ɪn'ventə]	Erfinder(in)
labour market ['leɪbə ˌmɑːkɪt]	Arbeitsmarkt
law of economics [ˌlɔːr‿əv‿iːkə'nɒmɪks]	Wirtschaftsgesetz
low interest rate [ˌləʊ‿'ɪntrəst‿ˌreɪt]	niedrige Zinsrate
Malaysia [mə'leɪʒə]	Malaysia
move [muːv]	Umzug
plant [plɑːnt]	Werk
to prosper ['prɒspə]	gedeihen
recession [rɪ'seʃn]	Flaute
to record [rɪ'kɔːd]	verzeichnen
to roar away [rɔːr‿ə'weɪ]	"wegdonnern"
sales [seɪlz]	Absätze
service industry ['sɜːvɪs ˌɪndʌstrɪ]	Dienstleistungsbranche
to shift [ʃɪft]	verschieben, verlagern
shopkeeper ['ʃɒpˌkiːpə]	Ladenbesitzer(in)
shopper ['ʃɒpə]	Käufer(in)
spending power ['spendɪŋ ˌpaʊə]	Kaufkraft
strike [straɪk]	Streik
to struggle with ['strʌgl wɪð]	kämpfen mit
to survive [sə'vaɪv]	überleben
to tend [tend]	neigen
to turn into [tɜːn‿'ɪntuː]	umwandeln in
two-speed [ˌtuː'spiːd]	mit zwei Geschwindigkeiten
unemployment rate [ˌʌnɪm'plɔɪmənt ˌreɪt]	Arbeitslosenquote
vacuum cleaner ['vækjuːm ˌkliːnə]	Staubsauger

D

access ['ækses]	Zugang, Zutritt
after all [ˌɑːftər‿'ɔːl]	schließlich
assistance [ə'sɪstəns]	Hilfe
back-up ['bækʌp]	Unterstützung
bright [braɪt]	hell, strahlend
car park ['kɑːpɑːk]	Parkplatz
council ['kaʊnsl]	(Stadt-, Grafschafts-)Rat
county ['kaʊntɪ]	Grafschaft, Bezirk
current ['kʌrənt]	aktuell, gegenwärtig
developer [dɪ'veləpə]	Entwickler(in), Bauunternehmer(in)
extract ['ekstrækt]	Auszug, Extrakt
facility [fə'sɪlətɪ]	Einrichtung, Möglichkeit
high class [ˌhaɪ'klɑːs]	hochwertig, erstklassig
housing ['haʊzɪŋ]	Unterkunft, Wohnungen
link [lɪŋk]	Verbindung
literature ['lɪtrətʃə]	Literatur, Informationsmaterial
location [ləʊ'keɪʃn]	Lage, Standort
out of town [ˌaʊt‿əv 'taʊn]	außerhalb der Stadt
permission [pə'mɪʃn]	Erlaubnis
phone call ['fəʊnkɔːl]	Anruf
promotion literature [prə'məʊʃn ˌlɪtrətʃə]	Werbematerial
racing bicycle ['reɪsɪŋ ˌbaɪsɪkl]	Rennrad
region ['riːdʒn]	Gebiet, Bezirk
to relocate [ˌriːləʊ'keɪt]	(Geschäftssitz) verlegen
to represent [ˌreprɪ'zent]	darstellen, stehen für
requirement [rɪ'kwaɪəmənt]	Bedarf, Erfordernis
shopping ['ʃɒpɪŋ]	Einkauf(en)

skilled [skɪld]	qualifiziert, ausgebildet
standard ['stændəd]	Niveau, Standard, Maßstab
up to [ʌp‿tuː]	bis zu

Unit 10

Starter

immigration [ˌɪmɪ'greɪʃn]	Einwanderung

A1

Amerasian [ˌæmər'eɪʒn]	amerikanisch-asiatisch
childhood ['tʃaɪldhʊd]	Kindheit
Chinese [tʃaɪ'niːz]	Chinese, Chinesin; chinesisch
cookie ['kʊkɪ]	Keks, Plätzchen
courage ['kʌrɪdʒ]	Mut
determination [dɪˌtɜːmɪ'neɪʃn]	Entschlossenheit, Bestimmung
down payment [ˌdaʊn‿'peɪmənt]	Anzahlung
draughtsman ['drɑːftsmən]	Zeichner(in)
to emigrate ['emɪgreɪt]	auswandern
emigration [ˌemɪ'greɪʃn]	Auswanderung
existence [ɪg'zɪstəns]	Existenz, Dasein
fashionable ['fæʃnəbl]	modisch
to flee [fliː]	fliehen, flüchten
to frighten ['fraɪtən]	erschrecken, Angst machen
to grow up [grəʊ‿'ʌp]	aufwachsen
happiness ['hæpɪnəs]	Glück, Heiterkeit, Zufriedenheit
hopeful ['həʊpfʊl]	zuversichtlich, vielversprechend
identity [aɪ'dentətɪ]	Identität, Übereinstimmung
to integrate ['ɪntɪgreɪt]	integrieren
to make matters worse [meɪk ˌmætəz‿'wɜːs]	zu allem Unglück
operator ['ɒpəreɪtə]	(Maschinen-)Bediener(in), Telefonvermittler(in)
original [ə'rɪdʒənl]	ursprünglich, originell
outsider [ˌaʊt'saɪdə]	Außenseiter(in)
possession [pə'zeʃn]	Besitz
quotation [kwəʊ'teɪʃn]	Zitat, Kostenvoranschlag
sceptical ['skeptɪkl]	skeptisch
soldier ['səʊldʒə]	Soldat(in)
title ['taɪtl]	Titel
Vietnamese [ˌvjetnə'miːz]	Vietnamese, Vietnamesin; vietnamesisch
visa ['viːzə]	Visum

A2

to arrest; arrest [ə'rest]	verhaften; Verhaftung
Asian ['eɪʒn]	Asiat(in); asiatisch
community [kə'mjuːnətɪ]	Bevölkerungsgruppe, Gemeinde
illegal [ɪ'liːgl]	illegal, gesetzlich unzulässig
Latin America [ˌlætɪn‿ə'merɪkə]	Lateinamerika
to loot [luːt]	plündern
Mexican ['meksɪkən]	Mexikaner(in); mexikanisch
Pole ['pəʊl]	Pole, Polin
Polish ['pəʊlɪʃ]	polnisch
riot ['raɪət]	Aufruhr, Krawall
tension ['tenʃn]	Spannung

C

circle ['sɜːkl]	Kreis, Kreislauf
to confirm [kən'fɜːm]	bestätigen
corn [kɔːn]	Getreide, Mais, Korn
draft [drɑːft]	Entwurf, Konzept
earnings ['ɜːnɪŋz]	Einkommen, Verdienst, Ertrag
economist [ɪ'kɒnəmɪst]	Wirtschaftswissenschaftler(in)
food stamp ['fuːd‿ˌstæmp]	Essensmarke
in the long run [ɪn‿ðə 'lɒŋ‿ˌrʌn]	auf lange Sicht
inflow ['ɪnfləʊ]	Zustrom
native-born [ˌneɪtɪv'bɔːn]	gebürtig
poll [pəʊl]	Umfrage, Abstimmung, Wahl
short-term ['ʃɔːttɜːm]	kurzfristig
society [sə'saɪətɪ]	Gesellschaft
structure ['strʌktʃə]	Struktur
theory ['θɪərɪ]	Theorie
unemployment benefit [ˌʌnɪm'plɔɪmənt ˌbenəfɪt]	Arbeitslosenunterstützung
version ['vɜːʃn]	Version, Fassung
welfare ['welfeə]	Wohl, Wohlfahrt, Sozialhilfe
word-by-word [ˌwɜːd‿baɪ‿'wɜːd]	Wort für Wort
worried ['wʌrɪd]	besorgt

D1

climate ['klaɪmət]	Klima
way of life [ˌweɪ‿əv‿'laɪf]	Lebensweise

D2

editor ['edɪtə]	Herausgeber(in), Redakteur(in)
to face (a problem) [feɪs / ə 'prɒbləm]	(einem Problem) gegenüberstehen
to publish ['pʌblɪʃ]	veröffentlichen

Unit 11

A

acceleration [ək͵selə'reɪʃn]	Beschleunigung
affordable [ə'fɔːdəbl]	erschwinglich
anti-lock brake [͵æntɪlɒk 'breɪk]	Antiblockierbremse
aspect ['æspekt]	Aspekt
assembly line [ə'semblɪ‿͵laɪn]	Fließband
automobile ['ɔːtəməbiːl]	Auto (AE)
based on ['beɪst‿ɒn]	basierend, gegründet auf
brake [breɪk]	Bremse
car maker ['kɑː͵meɪkə]	Automobilhersteller
carbon dioxide [͵kɑːbən‿daɪ'ɒksaɪd]	Kohlendioxyd
to challenge; challenge ['tʃælɪndʒ]	herausfordern; Herausforderung
combustion [kəm'bʌstʃn]	Verbrennung
concept ['kɒnsept]	Konzept, Entwurf
to consider [kən'sɪdə]	betrachten, denken (an), berücksichtigen
contradictory [͵kɒntrə'dɪktərɪ]	(sich) widersprechend, widersprüchlich
cruise control ['kruːz kən͵trəʊl]	Geschwindigkeits-kontrolle
crush zone ['krʌʃ‿͵zəʊn]	Knautschzone
to demonstrate ['demənstreɪt]	demonstrieren, zeigen
drawing board ['drɔːɪŋ ͵bɔːd]	Zeichenbrett
to evolve [ɪ'vɒlv]	(sich) entwickeln
flair [fleə]	Flair, Gespür, Talent, Begabung
forerunner ['fɔː͵rʌnə]	Vorläufer
guideline ['gaɪdlaɪn]	Richtlinie
hybrid ['haɪbrɪd]	Hybrid-, Misch-; Kreuzung, Mischung
hydrogen ['haɪdrədʒən]	Wasserstoff
in charge of [ɪn‿'tʃɑːdʒ‿əv]	verantwortlich für
to intend [ɪn'tend]	beabsichtigen
internal combustion engine [ɪn͵tɜːnl‿kəm'bʌstʃn‿͵endʒɪn]	Verbrennungsmotor
methane ['miːθeɪn]	Methan(-gas)
mobility [məʊ'bɪlətɪ]	Mobilität
rape seed ['reɪp ͵siːd]	Rapssamen
regulation [͵regjʊ'leɪʃn]	Vorschriften, Regelungen
seemingly ['siːmɪŋlɪ]	offensichtlich, scheinbar
side impact beam [͵saɪd 'ɪmpækt ͵biːm]	Seitenaufprallschutz
spirit ['spɪrɪt]	Geist(eshaltung), Sinn, Einstellung
to standardize ['stændədaɪz]	standardisieren
tough [tʌf]	streng, zäh
unimportant [͵ʌnɪm'pɔːtənt]	unwichtig

B

cheerful ['tʃɪəfʊl]	fröhlich, heiter
effort ['efət]	Anstrengung, Mühe, Versuch
to fall asleep [fɔːl‿ə'sliːp]	einschlafen
to ignore [ɪg'nɔː]	ignorieren, nicht beachten
in the first place [ɪn‿ðə 'fɜːst‿͵pleɪs]	zuerst
to promote [prə'məʊt]	(be)fördern, Werbung machen für
to recharge [͵riː'tʃɑːdʒ]	(wieder)aufladen
scrap heap ['skræp ͵hiːp]	Müll-, Schutt-, Schrotthaufen
zero ['zɪərəʊ]	null

C

to accompany [ə'kʌmpənɪ]	begleiten
battery ['bætərɪ]	Batterie
carburettor [kaːbjʊ'rɛtə]	Vergaser
coil [kɔɪl]	(Zünd-)Spule
combustion chamber [kəm'bʌstʃn‿͵tʃeɪmbə]	Brennkammer
to compress [kəm'pres]	komprimieren
compression [kəm'preʃn]	Kompression, Verdichtung
crankshaft ['kræŋkʃɑːft]	Kurbelwelle
cylinder ['sɪlɪndə]	Zylinder
damp [dæmp]	feucht
to distribute [dɪ'strɪbjuːt]	verteilen
distributor [dɪ'strɪbjʊtə]	Verteiler
four-stroke engine [͵fɔːstrəʊk‿'endʒɪn]	Viertaktmotor
to ignite [ɪg'naɪt]	zünden
ignition [ɪg'nɪʃn]	Zündung
intake ['ɪnteɪk]	Aufnahme, Ansaugen
lead [liːd]	Leitungskabel, Zuleitung
lifeless ['laɪfləs]	leblos
mixture ['mɪkstʃə]	Mischung
piston ['pɪstən]	Kolben
piston rod ['pɪstən ͵rɒd]	Pleuel-/Kolbenstange
silencer ['saɪlənsə]	Schalldämpfer
spark [spɑːk]	(Zünd-)Funke
spark plug ['spɑːkplʌg]	Zündkerze
stage [steɪdʒ]	Phase, Stadium, Bühne
stroke [strəʊk]	Schlag, Hub
to suck [sʌk]	(an)saugen
to turn over [tɜːn‿'əʊvə]	laufen (Motor)
voltage ['vəʊltɪdʒ]	elektrische Spannung

Unit 12

A

action ['ækʃn]	Maßnahme, Handlung

addition [ə'dɪʃn]	Hinzufügen, Ergänzung, Addition
ahead [ə'hed]	vor, voraus
air-conditioning ['eəkən,dɪʃənɪŋ]	Klimaanlage, Klimatisierung
anywhere ['enɪweə]	überall(hin), wo(hin) auch immer
appearance [ə'pɪərəns]	Erscheinung, Aussehen
to approve [ə'pruːv]	billigen, gutheißen
beauty ['bjuːtɪ]	Schönheit
blade [bleɪd]	Blatt, Schaufel, Klinge
coal-fired ['kəʊlfaɪəd]	mit Kohle betrieben
conventional [kən'venʃənl]	herkömmlich, konventionell
daffodil ['dæfədɪl]	Narzisse
decibel ['desɪbel]	Dezibel (Maß für Lautstärke)
to disturb [dɪ'stɜːb]	stören
double-glazed [,dʌbl'gleɪzd]	doppelt verglast
to eliminate [ɪ'lɪmɪneɪt]	beseitigen
gear [gɪə]	Getriebe, Gang
geo-thermal [,dʒiːəʊ'θɜːml]	geothermisch (Erdwärme nutzend)
goal [gəʊl]	Ziel, Tor
hydro-electric [,haɪdrəʊɪ'lektrɪk]	hydroelektrisch (Wasserkraft nutzend)
landscape ['lændskeɪp]	Landschaft
mechanism ['mekənɪzm]	Mechanismus
negative ['negətɪv]	negativ
nuclear ['njuːklɪə]	nuklear, Kern-, Atom-
off-shore ['ɒfʃɔː]	Off-shore, im Meer
output ['aʊtpʊt]	Produktion, Leistung, Output
positive ['pɒzətɪv]	positiv
potential [pə'tenʃl]	Potential, Möglichkeiten
to spoil [spɔɪl]	verderben, ruinieren
toxic ['tɒksɪk]	giftig
ugly ['ʌglɪ]	hässlich
unwilling [ʌn'wɪlɪŋ]	nicht bereit (etwas zu tun), widerwillig
visual ['vɪʒʊəl]	visuell, optisch
to wake (up) [weɪk‿'ʌp]	aufwachen, wecken
wave [weɪv]	Welle
Welsh [welʃ]	walisisch
windy ['wɪndɪ]	windig

B

anyway ['enɪweɪ]	sowieso, wie dem auch sei, jedenfalls
apparent [ə'pærənt]	offensichtlich, offenbar
critic ['krɪtɪk]	Kritiker(in)
fund [fʌnd]	Fond, Fundus, Mittel
furious ['fjʊərɪəs]	wütend, wild, heftig
remote [rɪ'məʊt]	entfernt, abgelegen

C

Electricity Board [ɪlek'tɪsətɪ ,bɔːd]	Elektrizitätsversorgungsunternehmen
heavy industry [,hevɪ‿'ɪndʌstrɪ]	Schwerindustrie
light industry [,laɪt‿'ɪndʌstrɪ]	Leichtindustrie
prime energy [,praɪm‿'enədʒɪ]	Primärenergie
pylon ['paɪlɒn]	Mast
sub-station ['sʌb,steɪʃn]	Umspannungswerk
transformer ['trænsfəmə]	Umwandler
transmission cable [trænz'mɪʃn ,keɪbl]	Leitungskabel
various ['veərɪəs]	verschieden, vielfältig

Unit 13

A

air-conditioner ['eəkən,dɪʃənə]	Klimaanlage
to ally; ally ['ælaɪ]	verbünden; Verbündete(r)
atmosphere ['ætməsfɪə]	Atmosphäre
attempt [ə'temt]	Versuch
aware [ə'weə]	bewußt
bather ['beɪðə]	Badende(r)
centigrade ['sentɪgreɪd]	Celsius
chemical ['kemɪkl]	chemisch
chlorofluorocarbon (CFC) [,klɔːrəʊ,flʊərəʊ'kɑːbən] [,siː ef'siː]	Fluorchlorkohlenwasserstoff (FCKW)
CO_2 [,siːəʊ'tuː]	Kohlendioxyd
to confront [kən'frʌnt]	gegenüberstellen, konfrontieren
confrontation [,kɒnfrʌn'teɪʃn]	Konfrontation
consequence ['kɒnsɪkwəns]	Folge, Konsequenz
to cut [kʌt]	reduzieren
dinosaur ['daɪnəsɔː]	Dinosaurier
emission [ɪ'mɪʃn]	Ausstoß, Emission
English Channel [,ɪŋglɪʃ‿'tʃænl]	Ärmelkanal
evolution [,iːvə'luːʃn]	Entwicklung, Evolution
exaggeration [ɪg,zædʒə'reɪʃn]	Übertreibung
extinction [ɪk'stɪŋkʃn]	Aussterben
flame [fleɪm]	Flamme
to flood [flʌd]	überfluten
foam [fəʊm]	Schaum
for better or for worse [fə‿,betər‿ɔː‿fə‿'wɜːs]	zum Besseren oder Schlechteren hin
forest ['fɒrɪst]	Wald
garbage ['gɑːbɪdʒ]	Abfall, Müll
greenhouse effect ['griːnhaʊs‿ɪ,fekt]	Treibhauseffekt, globale Erwärmung
to halt; halt [hɔːlt]	anhalten; Halt, Pause, Unterbrechung
harmful ['hɑːmfʊl]	schädlich
Holy Bible [,həʊlɪ 'baɪbl]	(Heilige) Bibel
huge [hjuːdʒ]	riesig

humanity [hjuːˈmænətɪ]	Menschheit, Menschlichkeit
immune system [ɪˈmjuːn ˌsɪstəm]	Immunsystem
infertile [ɪnˈfɜːtaɪl]	unfruchtbar
irreplaceable [ˌɪrɪˈpleɪsəbl]	unersetzlich
lack [læk]	Mangel
massive [ˈmæsɪv]	massiv, wuchtig, enorm
medical [ˈmedɪkl]	ärztlich, medizinisch
Mediterranean (Sea) [ˌmedɪtəˈreɪnɪən]	Mittelmeer
migration [maɪˈgreɪʃn]	(Ab-)Wanderung
mobilization [ˌməʊbɪlaɪˈzeɪʃn]	Mobilisierung
ordinary [ˈɔːdɪnrɪ]	gewöhnlich
ozone layer [ˈəʊzəʊn‿ˌleɪə]	Ozonschicht
to pass away [pɑːs‿əˈweɪ]	entschlafen, hinscheiden; Umschreibung für: sterben
to pay attention [peɪ‿əˈtenʃn]	aufmerksam sein
pesticide [ˈpestɪsaɪd]	Pestizid
to poison; poison [ˈpɔɪzən]	vergiften; Gift
to release [rɪˈliːs]	freilassen, veröffentlichen
to ruin [ˈrʊɪn]	ruinieren
to sacrifice; sacrifice [ˈsækrɪfaɪs]	opfern; Opfer
scale [skeɪl]	Maß(-stab)
seal [siːl]	Seehund
sewage [ˈsuːɪdʒ]	Abwasser
smokestack [ˈsməʊkstæk]	Schornstein
species [ˈspiːsiːz]	(Tier-, Pflanzen-)Art, Spezies
stream [striːm]	Bach, Wasserlauf
survival [səˈvaɪvl]	Überleben
threat [θret]	(Be-)Drohung
to threaten [θretn]	(be-)drohen
treatment [ˈtriːtmənt]	Behandlung
tropical [ˈtrɒpɪkl]	tropisch
uninhabitable [ˌʌnɪnˈhæbɪtəbl]	unbewohnbar
unless [ʌnˈles]	wenn ... nicht, es sei denn
wartime [ˈwɔːtaɪm]	Kriegszeit
to weaken [ˈwiːkən]	schwächen

B

to convert [kənˈvɜːt]	umwandeln
to establish [ɪˈstæblɪʃ]	schaffen, gründen, errichten
front-page [ˌfrʌnt‿ˈpeɪdʒ]	Titelseite
impressive [ɪmˈpresɪv]	beeindruckend
NASA (National Aeronautics and Space Administration) [ˈnæsə] [ˈnɑːzə] [ˌnæʃənəl ˌeərəˈnɔːtɪks‿ənd ˈspeɪs‿ədˌmɪnɪstreɪʃn]	Raumfahrtbehörde
power plant [ˈpaʊə ˌplɑːnt]	Kraftwerk
to turn into [tɜːn‿ˈɪntuː]	umwandeln in
waste-melting [ˈweɪst‿ˌmeltɪŋ]	"Einschmelzen" von Abfall

C

avoidance [əˈvɔɪdəns]	Vermeidung
cardboard packaging [ˌkɑːdbɔːd ˈpækɪdʒɪŋ]	Kartonverpackung
incineration [ɪnˌsɪnəˈreɪʃn]	Verbrennung
landfill deposit [ˈlændfɪl dɪˌpɒzɪt]	Mülldeponie
out-of-date [ˌaʊt‿əv ˈdeɪt]	veraltet, überholt, altmodisch
physical [ˈfɪzɪkl]	physikalisch, physisch
to recharge [ˌriːˈtʃɑːdʒ]	wieder aufladen
to replace [rɪˈpleɪs]	ersetzen
soil [sɔɪl]	Boden, Erde
substitution [ˌsʌbstɪˈtjuːʃn]	Ersetzen, Austausch
to treat [triːt]	behandeln
TV set [ˌtiːˈviː ˌset]	Fernsehgerät

Unit 14

A

accuracy [ˈækjʊrəsɪ]	Genauigkeit
aeroplane [ˈeərəpleɪn]	Flugzeug
airliner [ˈeəlaɪnə]	Verkehrsflugzeug
brain [breɪn]	Gehirn
brief [briːf]	kurz, knapp, gering
button [ˈbʌtn]	Knopf
concurrent [ˌkənˈkʌrənt]	gleichzeitig
extreme [ɪkˈstriːm]	extrem, äußerst, höchst
to fit together [fɪt‿təˈgeðə]	zusammenpassen
numerical [njuːˈmerɪkl]	numerisch, zahlenmäßig
to oversee [ˌəʊvəˈsiː]	überwachen, leiten
perfect [ˈpɜːfɪkt]	vollkommen, perfekt
prototype [ˈprəʊtəʊtaɪp]	Prototyp
to roll [rəʊl]	rollen
to shorten [ˈʃɔːtn]	kürzen
spectacular [spekˈtækjələ]	spektakulär
thousandth [ˈθaʊzənθ]	Tausendstel
to tie [taɪ]	binden, bündeln
twin-engined [ˌtwɪn‿ˈendʒɪnd]	zweimotorig

B

common ['kɒmən]	gemeinsam
considerable [kən'sɪdərəbl]	beträchtlich
solid ['sɒlɪd]	massiv, verläßlich, ganz
sophisticated [sə'fɪstɪkeɪtɪd]	hochentwickelt, anspruchsvoll, ausgeklügelt

C

acceptable [ək'septəbl]	annehmbar, akzeptabel
central ['sentrəl]	zentral
continuous [kən'tɪnjʊəs]	ununterbrochen, fortlaufend
to display [dɪ'spleɪ]	zeigen
DVD drive [ˌdiːviː'diː ˌdraɪv]	DVD-Laufwerk
enormous [ɪ'nɔːməs]	enorm, riesig, gewaltig
headset ['hedset]	Kopfhörer
to implement ['ɪmpləment]	erfüllen, umsetzen, ausführen
to limit ['lɪmɪt]	begrenzen, beschränken
scanner ['skænə]	Scanner, Abtaster

Unit 15

A

to affect [ə'fekt]	sich auswirken auf, betreffen
anchorage ['æŋkərɪdʒ]	Anker, Verankerung
to attach [ə'tætʃ]	befestigen
beam [biːm]	Balken, Träger, Stütze
belt [belt]	Gürtel, Gurt
to bend [bend]	(sich) biegen
builder ['bɪldə]	Erbauer(in), Bauunter-nehmer(in)
building site ['bɪldɪŋ ˌsaɪt]	Baustelle
civil engineer [ˌsɪvl‿ˌendʒɪ'nɪə]	Bauingenieur(in)
cofferdam ['kɒfədæm]	Kastendamm, Kofferdamm
concrete ['kɒŋkriːt]	Beton
consequently ['kɒnsɪkwəntlɪ]	daraus folgend, infolgedessen
to conserve [kən'sɜːv]	erhalten, schonen
to construct [kən'strʌkt]	bauen, konstruieren
construction [kən'strʌkʃn]	Bau, Konstruktion
earthquake ['ɜːθkweɪk]	Erdbeben
to erect [ɪ'rekt]	errichten, aufbauen
to face [feɪs]	gegenüberstehen
foundation [faʊn'deɪʃn]	Gründung, Fundament
helmet ['helmɪt]	Helm
in case of [ɪn‿'keɪs‿əv]	im Falle von
itself [ɪt'self]	selbst
lane [leɪn]	(Fahr-)spur, Landstraße, Weg
misty ['mɪstɪ]	neblig, dunstig
net [net]	Netz
to pump; pump [pʌmp]	pumpen; Pumpe
rope [rəʊp]	Seil
to span; span [spæn]	spannen; Spanne
suspender [sə'spendə]	Aufhängevorrichtung, Hosenträger
suspension [sə'spenʃn]	Aufhängung, Ausschluss
suspension bridge [sə'spenʃn ˌbrɪdʒ]	Hängebrücke
tides [taɪdz]	Gezeiten
unsupported [ˌʌnsə'pɔːtɪd]	ungestützt, nicht unterstützt

B

besides [bɪ'saɪdz]	außerdem
to bother ['bɒðə]	belästigen, beschäftigen, Sorgen machen
bracket ['brækɪt]	Klammer
to collapse [kə'læps]	zusammenbrechen
compact ['kɒmpækt]	kompakt, komprimiert
historical [hɪ'stɒrɪkl]	historisch
multi-storey car park [ˌmʌltɪstɔːrɪ 'kɑːpɑːk]	Parkhaus
objective [ɒb'dʒektɪv]	Ziel
pedestrian [pə'destrɪən]	Fußgänger(in)
pedestrian mall [pə'destrɪən ˌmɔːl]	Fußgängerzone
pile [paɪl]	Pfeiler, Pfahl, Haufen
to plot [plɒt]	zeichnen, abstecken
preparation [ˌprepə'reɪʃn]	Vorbereitung
to rot [rɒt]	verrotten
to sink [sɪŋk]	sinken
thorough ['θʌrə]	gründlich
wooden ['wʊdn]	hölzern

C

cemetery ['semətrɪ]	Friedhof
flyover ['flaɪˌəʊvə]	Überführung
pavement ['p eɪvmənt]	(BE) Bürgersteig, (AE) Straße
valley ['vælɪ]	Tal

Unit 16

to bark [bɑːk]	bellen
to check on sth. ['tʃek‿ɒn 'sʌmθɪŋ]	nach etwas sehen
cool box ['kuːlbɒks]	Eisfach
to download ['daʊnləʊd]	herunterladen (z.B. Daten, Musik)
file [faɪl]	Akte, Datei
interconnected [ˌɪntəkə'nektɪd]	vernetzt
interconnectivity [ˌɪntəkɒnek'tɪvətɪ]	Vernetzung
kettle ['ketl]	Wasserkessel

to link up [lɪŋk‿'ʌp]	anschließen
to operate ['ɒpəreɪt]	bedienen
recipe ['resɪpɪ]	Rezept
remote control [rɪˌməʊt‿kən'trəʊl]	Fernbedienung
remotely [rɪ'məʊtlɪ]	fern, entfernt
short-range radio signal [ˌʃɔ:treɪdʒ‿'reɪdɪəʊ ˌsɪgnl]	Kurzwelle
sink [sɪŋk]	Waschbecken
smart home ['smɑ:t ˌhəʊm]	intelligentes Heim
sufficient [sə'fɪʃnt]	ausreichend
to transmit [trænz'mɪt]	senden
to unlock [ʌn'lɒk]	aufschließen
wireless ['waɪələs]	drahtlos

Unit 17

to account for [ə'kaʊnt fɔ:]	ausmachen
to affect [ə'fekt]	sich auswirken auf
build-up ['bɪldʌp]	Zunahme
to compete for sth. [kəm'pi:t fə ˌsʌmθɪŋ]	um etwas wetteifern, kämpfen
contaminated [kən'tæmɪneɪtɪd]	verschmutzt, verunreinigt, verseucht
crops [krɒps]	Feldfrüchte
disease [dɪ'zi:z]	Krankheit
to experience [ɪk'spɪərɪəns]	erleben, erfahren
to extract [ɪk'strækt]	gewinnen, entnehmen
farmland ['fɑ:mlənd]	Ackerland
freshwater ['freʃˌwɔ:tə]	Süßwasser
to grow [grəʊ]	anbauen
infertile [ɪn'fɜ:taɪl]	unfruchtbar
irrigation [ˌɪrɪ'geɪʃn]	Bewässerung
mangrove swamp ['mæŋgrəʊv ˌswɒmp]	Mangrovensumpf
raw sewage [ˌrɔ: 'su:ɪdʒ]	ungeklärte Abwässer
to rely on [rɪ'laɪ‿ɒn]	abhängig sein von
rice [raɪs]	Reis
rural ['rʊərəl]	ländlich
scarcity ['skeəsətɪ]	Knappheit
sewage ['su:ɪdʒ]	Abwasser
shortage ['ʃɔ:tɪdʒ]	Mangel, Knappheit
soil [sɔɪl]	Boden
urban ['ɜ:bən]	städtisch, Stadt-
water table ['wɔ:tə ˌteɪbl]	Grundwasserspiegel
worrying ['wʌrɪɪŋ]	besorgniserregend

Unit 18

accidental [ˌæksɪ'dentl]	zufällig
architect ['ɑ:kɪtekt]	Architekt(in)
background ['bækgraʊnd]	hier: Herkunft, Verhältnisse
brick [brɪk]	Ziegelstein
bunker ['bʌŋkə]	Bunker
to collapse [kə'læps]	einstürzen
concrete ['kɒŋkri:t]	Beton
to counteract [ˌkaʊntər'ækt]	entgegenwirken
crush-proof ['krʌʃpru:f]	einsturzsicher
destructive [dɪ'strʌktɪv]	zerstörerisch, zerstörend
destruction [dɪ'strʌkʃn]	Zerstörung
elevator ['eləveɪtə]	Fahrstuhl, Aufzug
entire [ɪn'taɪə]	ganz, komplett
fire-proof ['faɪəpru:f]	feuersicher
(steel-)frame [ˌsti:l 'freɪm]	Rahmen (aus Stahl)
gallon ['gælən]	Gallone
to gather ['gæðə]	sich versammeln
to get rid of [get 'rɪd‿əv]	loswerden
horrified ['hɒrɪfaɪd]	entsetzt
to implode [ɪm'pləʊd]	implodieren
incredible [ɪn'kredɪbl]	unglaublich
to live out (dream) [lɪv‿'aʊt]	(Traum) verwirklichen
to maintain [meɪn'teɪn]	behaupten
skyscraper ['skaɪˌskreɪpə]	Wolkenkratzer
to spread [spred]	(sich) verbreiten
steel support [ˌsti:l sə'pɔ:t]	Stahlträger
technique [tek'ni:k]	Technik
twin towers [ˌtwɪn 'taʊəz]	Zwillingstürme
upper ['ʌpə]	obere(r,s)
urbanisation [ˌɜ:bənaɪ'zeɪʃn]	Verstädterung, Urbanisierung

Unit 19

alien ['eɪlɪən]	fremd
bioengineering [ˌbaɪəʊendʒɪ'nɪərɪŋ]	Biotechnik
biotechnology [ˌbaɪəʊtek'nɒlədʒɪ]	Biotechnologie
blood clotting ['blʌd ˌklɒtɪŋ]	Bildung von Blutgerinnseln
capsule ['kæpsju:l]	Kapsel
clone [kləʊn]	Klon
compound ['kɒmpaʊnd]	(chem.) Verbindung
to contaminate [kən'tæmɪneɪt]	verunreinigen, vergiften
to crush [krʌʃ]	zerdrücken, zerquetschen
disease [dɪ'zi:z]	Krankheit
drug [drʌg]	Arznei
to dry [draɪ]	trocknen
to endanger [ɪn'deɪndʒə]	gefährden
fried [fraɪd]	gebraten
genetically modified [dʒəˌnetɪklɪ 'mɒdɪfaɪd]	gentechnisch verändert
genetic research [dʒəˌnetɪk 'ri:sɜ:tʃ]	Genforschung
goat [gəʊt]	Ziege
to implant [ɪm'plɑ:nt]	einpflanzen, implantieren
to inject [ɪn'dʒekt]	(ein)spritzen, injizieren
lab [læb]	Labor

mashed (potatoes) [ˌmæʃt‿pə'teɪtəʊz]	zerdrückt (Kartoffelbrei)
mammal ['mæml]	Säugetier
medicine cabinet ['medsɪn ˌkæbɪnət]	Arzneimittelschrank
to milk [mɪlk]	melken
to modify ['mɒdɪfaɪ]	verändern, modifizieren
pace [peɪs]	Tempo, Schritt
pharmaceutical [ˌfɑːmə'sjuːtɪkl]	pharmazeutisch
pharming ['fɑːmɪŋ]	(Kunstwort) Landwirtschaft betreiben zu pharmazeutischen Zwecken
pill [pɪl]	Pille
protein ['prəʊtiːn]	Protein
rapid ['ræpɪd]	schnell
to refrigerate [rɪ'frɪdʒəreɪt]	im Kühlschrank aufbewahren, kühlen
research ['riːsɜːtʃ]	Forschung
researcher ['riːsɜːtʃə]	Forscher(in)
sensitive ['sensətɪv]	empfindlich, heikel
surgery ['sɜːdʒərɪ]	Chirurgie
to swallow ['swɒləʊ]	schlucken
transgenic [ˌtrænz'dʒenɪk]	Gene von anderen Lebewesen beinhaltend
vaccine ['væksiːn]	Impfstoff
prediction [prɪ'dɪkʃn]	Vorhersage
quantum ['kwɒntəm]	Quantum
race [reɪs]	Rasse
radical ['rædɪkl]	radikal
recognition [ˌrekəg'nɪʃn]	(Wieder)erkennung
relationship [rɪ'leɪʃnʃɪp]	Verhältnis, Beziehung
routinely [ruː'tiːnlɪ]	routinemäßig
scientific [ˌsaɪən'tɪfɪk]	wissenschaftlich
silicon chip ['sɪlɪkən ˌtʃɪp]	Siliziumchip
slope [sləʊp]	Gefälle, Schräge, Abhang
stock market ['stɒk ˌmɑːkɪt]	Börse
storage ['stɔːrɪdʒ]	Lagerung, Speicherung
trading ['treɪdɪŋ]	Handel
trial and error [ˌtraɪl‿n‿'erə]	Ausprobieren
two-legged [ˌtuː'legd] [ˌtuː'legɪd]	zweibeinig
unbelievable [ˌʌnbɪ'liːvəbl]	unglaublich
to wire up [waɪər‿'ʌp]	anschließen, verkabeln
wisdom ['wɪzdəm]	Weisheit, Kenntnisse

Unit 20

to adjust [ə'dʒʌst]	anpassen
advance [əd'vɑːns]	Fortschritt
automaton [ɔː'tɒmətən]	Automat, sich selbst bewegender Kunstmensch
bio-chips ['baɪəʊtʃɪps]	Chips aus lebender Materie
childlike ['tʃaɪldlaɪk]	wie ein Kind
common sense [ˌkɒmən 'sens]	gesunder Menschenverstand
DNA sequencing [ˌdiːen'eɪ ˌsiːkwənsɪŋ]	DNA-Sequenzbildung (-Aufeinanderfolge)
humanoid ['hjuːmənɔɪd]	menschenartig
to lay the table [ˌleɪ ðə 'teɪbl]	den Tisch decken
limit ['lɪmɪt]	Grenze
to manipulate [mə'nɪpjəleɪt]	manipulieren
membrane ['membreɪn]	Membrane
microprocessor ['maɪkrəʊˌprəʊsesə]	Mikroprozessor
molecular bond [məˌlekjələ 'bɒnd]	molekulare Verbindung
non-threatening [ˌnɒn'θretnɪŋ]	harmlos, nicht bedrohlich
optical ['ɒptɪkl]	optisch
posture control ['pɒstʃə kənˌtrəʊl]	etwa: Körperhaltungsregler

Word list (alphabetical)

AE = American English
BE = British English

A

a bit	ein bisschen
*a couple of**	*ein paar*
a little	*etwas, ein bisschen*
a lot of	*viel*
a number of	*einige*
ability	Fähigkeit
to be able	*können*
to be about	sich um etwas handeln
above	oben
abroad	*im/ins Ausland*
acceleration	Beschleunigung
to accept	*annehmen, akzeptieren*
acceptable	annehmbar, akzeptabel
access	Zugang, Zutritt
accident	*Unfall*
accidental	zufällig
accommodation	Unterkunft, Zimmer
to accompany	begleiten
according to (him)/(her)	(seiner)/(ihrer) Meinung nach, laut
to account for	ausmachen, erklären
accountancy	Rechnungswesen
accuracy	Genauigkeit
accurate	*genau*
to accuse	beschuldigen, anklagen
to ache; ache	schmerzen; Schmerz
to achieve	erreichen, erzielen
acid rain	saurer Regen
to acquire	sich aneignen
to act	handeln
action	Maßnahme, Handlung
active	aktiv
activity	*Aktivität*
actual	tatsächlich
to adapt	bearbeiten
to add	*hinzufügen, addieren*
addictive	süchtig machend
addition	Hinzufügen, Ergänzung, Addition
additional	zusätzlich
address	*Adresse*
adjective	Adjektiv
to adjust	einstellen, anpassen
adult	Erwachsene(r)
advance	Fortschritt
advanced	fortgeschritten
advantage	*Vorteil*
adventure	Abenteuer
advert	Kurzform von „advertisement" = Werbeanzeige
to advertise	*werben, inserieren*
advertisement	*(Werbe-)Anzeige*
advertiser	Inserent(in), Auftraggeber(in) von Werbesendungen
advertising	*Werbung*
advertising campaign	Werbekampagne
advertising manager	Werbeleiter(in)
advice	*Rat(schlag)*
to advise	*(be)raten*
aeroplane	Flugzeug
to affect	sich auswirken auf, betreffen
affinity	Verbundenheit, Verwandtschaft
to afford	sich leisten
affordable	erschwinglich
to be afraid (of)	*Angst haben (vor)*
after all	schließlich
afternoon	*Nachmittag*
afterwards	*danach*
again	*wieder*
again and again	immer wieder
against	*gegen*
age	*(Zeit-)Alter*
aged …	… Jahre alt
aggressive	aggressiv
to agree	*einverstanden sein, zustimmen*
agreement	*Übereinstimmung, Übereinkunft*
agricultural	landwirtschaftlich
ahead	vor, voraus
aid	Hilfe, Hilfsmittel
to aim (at); aim	zielen (auf); Ziel
air	*Luft*
air-conditioner	Klimaanlage
air-conditioning	Klimaanlage, Klimatisierung
airliner	Verkehrsflugzeug
airport	*Flughafen*
airtime	Sendezeit
alien	fremd
alive	lebend, lebendig, aktiv
all	*alle(s), ganz*
all but	fast
all kind(s) of	alle Arten von, alle möglichen
all over the world	in der ganzen Welt
all-inclusive	alles beinhaltend
to allow	*erlauben*
to ally; ally	verbünden; Verbündete(r)
almost	*fast, beinahe*
alone	*allein*
along	*entlang*
already	*schon*
also	*auch*
alternative	*Alternative, Wahl; alternativ*
although	*obwohl*
altogether	insgesamt, völlig
always	*immer*
ambition	Ehrgeiz, Ambition
Amerasian	amerikanisch-asiatisch
among	*zwischen, unter*
amount	*Betrag, Summe*
anchorage	Anker, Verankerung
anger	Zorn, Wut
angry	*wütend*
animal	*Tier*
to announce	*bekannt geben, ansagen*
announcement	*Bekanntgabe, Durchsage*
annual output	Jahresproduktion
another	*noch eine(r), ein(e) andere(r)*
to answer; answer	*(be)antworten; Antwort*
anti-lock brake	Antiblockierbremse
antonym	Antonym, Wort mit gegensätzlicher Bedeutung
any	*irgendeine(r,s)*
anytime	*jederzeit*
anybody	*irgendjemand*
anyone	*irgendjemand*
anything	*irgendetwas*
anyway	sowieso, wie dem auch sei, jedenfalls
anywhere	überall(hin), wo(hin) auch immer
to apologize	sich entschuldigen
apparatus	Gerät
apparent	offensichtlich, offenbar
to appear	*(er)scheinen*
appearance	Erscheinung, Aussehen
apple	*Apfel*
appliance	Gerät
applicant	Bewerber(in)
application	Anwendung
to apply (for)	*sich bewerben (um)*
to appoint	ernennen, bestimmen
appointment	Verabredung, Termin, Ernennung
approach	Ansatz, Annäherung
to approve	billigen, gutheißen
approval	Zustimmung
architect	Architekt(in)
archbishop	Erzbischof
area	*Gebiet*
to argue	argumentieren, behaupten

*(*Der als bekannt vorausgesetzte Grundwortschatz ist kursiv gedruckt.)*

arms	Waffen
around	*herum, um, ungefähr*
to arrange	aufstellen, planen, veranlassen
arrangement	Vereinbarung, Arrangement
to arrest; arrest	verhaften; Verhaftung
arrival	*Ankunft*
to arrive	*ankommen*
article	*Artikel*
as	*wie, da, als*
as a result	folglich
as well	*auch, ebensogut*
as well as	sowie
ash	Asche
Asian	Asiat(in); asiatisch
to ask (for)	*fragen, bitten (um)*
aspect	Aspekt
to assemble	zusammenbauen
assembly line	Fließband
assertiveness	Selbstbewusstsein
asset	Vermögen(-swert), Besitz
assistance	Hilfe
assistant	Assistent(in)
to associate	assoziieren, in Verbindung bringen
at all costs	um jeden Preis
at first	*anfangs*
at home	*zu Hause*
at last	*schließlich, endlich*
at least	*wenigstens*
at once	*sofort*
at present	*im Augenblick, jetzt*
at the moment	*jetzt, im Moment*
at work	*bei der Arbeit*
atmosphere	Atmosphäre
to attach	befestigen
to attack	in Angriff nehmen, angreifen
attempt	Versuch
to attend	*(Schule) besuchen*
attendance	Besuch
attention	*Aufmerksamkeit*
attitude	Haltung, Einstellung
attraction	Anziehungskraft
attractive	reizvoll, anziehend, attraktiv
audio	Audio, Hör-
aunt	*Tante*
Austria	Österreich
author	Autor(in), Verfasser(in)
to automate	automatisieren
automatic	automatisch
automaton	Automat, sich selbst bewegender Kunstmensch
automobile	Auto (Am.)
autumn	*Herbst*
availability	Vorhandensein, Verfügbarkeit
available	verfügbar, vorhanden
average	Durchschnitt
to avoid	vermeiden
avoidance	Vermeidung
aware	bewusst
away	*weg*
awful	*furchtbar, schrecklich*

B

Bachelor of Arts (B.A.)	niedrigster akad. Grad
back	*Rücken; zurück*
back-up	Unterstützung
background	Hintergrund, Herkunft, Verhältnisse
bad	*schlecht, schlimm, böse*
bag	*Tüte, Tasche*
baggage	*Gepäck*
bagless	hier: ohne Staubsaugerbeutel
to bake	*backen*
baker	*Bäcker(in)*
balcony	*Balkon*
Baltic	baltisch; Ostsee
bank	*Bank (Geldinstitut)*
bank holiday	öffentl. Feiertag
bar graph	Säulendiagramm
to bark	bellen
base	Basis, Stützpunkt, Ausgangspunkt
based on	basierend, gegründet auf
basic	Grund-; grundlegend
basket	*Korb*
bather	Badende(r)
bathroom	*Badezimmer*
battery	Batterie
beach	*Strand*
beam	Balken, Träger, Stütze
beard	*Bart*
beautiful	*schön*
beauty	Schönheit
to become, (became, become)	*werden*
because	*weil*
because of	*wegen*
bed	*Bett*
bedsitter	Einzimmerwohnung
bedroom	*Schlafzimmer*
beer	Bier
before	*vorher, ehe*
to begin, (began, begun)	*beginnen*
beginner	Anfänger(in)
beginning	*Anfang*
behaviour	Benehmen
behind	*hinter*
Belgian	Belgier(in); belgisch
Belgium	Belgien
to believe	*glauben*
to belong to	*gehören*
belt	Gürtel, Gurt
to bend	(sich) biegen
benefit	Zuwendung, soziale Leistung
besides	außerdem
best	*am besten*

best wishes	*alles Gute*
best known	bekannteste(r,s)
better	*besser*
between	*zwischen*
bicycle	*Fahrrad*
big	*groß*
bike	*Fahrrad (Kurzform)*
bill	*Rechnung*
billion	Billion, (AE: Milliarde)
bio-chips	Chips aus lebender Materie
bioengineering	Biotechnik
biomass	Biomasse
biotechnology	Biotechnologie
bird	*Vogel*
birth	*Geburt*
birthday	*Geburtstag*
biscuit	Keks
bit	*Stückchen*
to bite, (bit, bitten)	*beißen*
black	*schwarz*
blackboard	*(Wand-)Tafel*
blade	Blatt, Schaufel, Klinge
to blame (for)	*verantwortlich machen (für)*
to blink	blinken
block of flats	*Wohnblock*
blood	*Blut*
blood clotting	Bildung von Blutgerinnseln
blue	*blau*
board	*Brett*
boat	*Boot*
body	*Körper*
bog	Moor
to boil	*kochen*
bold type	Fettdruck
bone	Knochen
to book; book	*buchen; Buch*
booking	Buchung
bookshop	Buchhandlung
boot	*Stiefel*
border	*Grenze*
bored	*gelangweilt*
boring	*langweilig*
to borrow	*sich ausleihen*
both	*beide*
to bother	belästigen, beschäftigen, sich Sorgen machen
bottle	*Flasche*
bottom	*Boden*
(to be) bound (to do sth.)	(etwas) bestimmt (tun), verpflichtet, unterwegs
boundary	Grenze
box	*Schachtel, Karton, Kiste*
boy	*Junge*
to boycott; boycott	boykottieren; Boykott
boyfriend	*Freund*
bracket	Klammer
brain	Gehirn
brake	Bremse
brand name	Markenname

bread	*Brot*
to break, (broke, broken); break	*(zer)brechen; Pause*
to break down	*zusammenbrechen*
breakfast	*Frühstück*
brick	Ziegelstein
bridge	*Brücke*
brief	kurz, knapp, gering
bright	hell, strahlend
to bring, (brought, brought)	*bringen*
broadband	Breitband
to broadcast, (broadcast, broadcast)	*übertragen, senden*
to broaden the mind	den Horizont erweitern
brochure	*Broschüre, Prospekt*
brother	*Bruder*
brown	*braun*
to brush; brush	*bürsten; Bürste*
BTEC National Diploma	etwa: Fachhochschulreifeprüfung
BTU British Thermal Unit	BTU = Brit. Wärmeeinheit
bubble	Blase
to build, (built, built)	*bauen*
builder	Erbauer(in), Bauunternehmer(in)
building	*Gebäude*
building site	Baustelle
build-up	Zunahme
bulb	(Glüh)birne
bumper	Stoßstange
bunker	Bunker
bureaucracy	Bürokratie
to burn, (burnt, burnt)	*brennen*
bus pass	Busausweis
business	*Geschäft*
business studies	etwa: Fachbereich Wirtschaft
busy	*beschäftigt*
but	*aber*
butcher	*Metzger(in)*
button	Knopf
to buy, (bought, bought)	*kaufen*
buyer	*Käufer(in)*
by	*durch, bei, neben, bis*
by far	bei weitem
by rail	mit der Bahn
by the way	*übrigens*
by-product	Nebenprodukt

C

cabinet	Kabinett, Schrank
cable	Kabel
cable TV	Kabelfernsehen
CAD (Computer Aided Design)	Computerunterstützte Zeichnung (Entwurf)
cake	*Kuchen*
to calculate	*(be)rechnen*
to call; call	*(an)rufen, nennen; Anruf*
call centre	Callcenter, telefon. Auskunftsdienst
to call up	anrufen
CAM (Computer Aided Manufacturing)	Computerunterstützte Fertigung
camera	*Kamera*
to camp; camp	*zelten, lagern; Lager*
campaign	Kampagne, Werbefeldzug
campus	Campus, Universitätsgelände
can	*Büchse, Dose; können*
can't stand	nicht ausstehen können
candidate	Bewerber(in)
canteen	Kantine
capacity	Kapazität, Leistungsfähigkeit
capital	*Großbuchstabe, Hauptstadt*
capsule	Kapsel
car	*Auto*
car body	Karosserie
car maker	Automobilhersteller
car park	Parkplatz
carbon dioxide	Kohlendioxyd
carburettor	Vergaser
card	*Karte*
cardboard packaging	Kartonverpackung
cardiologist	Herzspezialist(in)
to care; care	*sorgen; Sorge*
career	*Beruf, Laufbahn, Karriere*
careful	*sorgfältig*
careless	*sorglos*
cargo	*Fracht, Ladung*
to carry	*tragen*
to carry out	ausführen
cartridge	Kassette, Patrone
case	*Fall, Koffer, Kiste*
cash	*Bargeld*
cassette	Kassette
cat	*Katze*
to catch, (caught, caught)	*fangen*
to cause; cause	*verursachen; Ursache*
ceiling	*(Zimmer-)Decke*
cell	Zelle
cemetery	Friedhof
centigrade	Celsius
central	zentral
Central Electricity Generating Board (CEGB)	etwa: zentrale Energieversorgung (England/Wales)
centre	*Zentrum, Mittelpunkt*
century	*Jahrhundert*
certain	*sicher*
certificate	Zeugnis
CFC (chlorofluorocarbon)	FCKW (Fluorchlorkohlenwasserstoff)
chain	Kette
chair	*Stuhl*
chalk	*Kreide*
to challenge; challenge	herausfordern; Herausforderung
to change; change	*wechseln, (sich) (ver)ändern; Wechsel(-geld), Änderung*
channel	*Kanal*
chart	*Schaubild, Tabelle, Karte*
cheap	*billig*
to check; check	*kontrollieren; Kontrolle*
cheerful	fröhlich, heiter
cheers	*Tschüs!, Prost!*
cheese	*Käse*
chemical	chemisch
chemist	Chemiker(in)
child	*Kind*
childcare assistant	Kinderpfleger(in)
childhood	Kindheit
childlike	wie ein Kind
China	China
Chinese	Chinese, Chinesin; chinesisch
chips	*Pommes Frites, Kartoffelchips*
chocolate	*Schokolade, Praline*
choice	*Wahl*
to choose, (chose, chosen)	*(aus)wählen*
Christmas	*Weihnachten*
church	*Kirche*
cigarette	*Zigarette*
CIM (Computer Integrated Manufacturing)	Computerintegrierte Fertigung
cinema	*Kino*
circle	Kreis, Kreislauf
citizen	Bürger(in)
city	*Stadt*
civil engineer	Bauingenieur(in)
civil engineering	Bauwesen, Hoch- und Tiefbau
civilized	zivilisiert
to claim; claim	behaupten, beanspruchen; Anspruch
class	*Klasse*
classmate	*Klassenkamerad(in)*
classroom	*Klassenzimmer*
to clean; clean	*reinigen; sauber*
to clear; clear	*säubern, räumen; klar*
clever	*klug*
climate	Klima
to climb	*klettern*
clock	*Uhr*
to clone; clone	klonen; Klon
to close; close	*schließen; dicht, nahe*
to close down	schließen, zumachen (Betrieb, Firma)
clothes	*Kleidung*
cloud	*Wolke*

CO_2	Kohlendioxyd
coach	Trainer(in)
coaching days	Postkutschenzeit
coal	*Kohle*
coal-fired	mit Kohle betrieben
coast	*Küste*
coastal	Küsten-
coat	*Mantel*
to code; code	kodieren; Kode
coffee	*Kaffee*
cofferdam	Kastendamm, Kofferdamm
coil	(Zünd-)Spule
coin	*Münze*
cold	*kalt*
to counteract	entgegenwirken
to collapse	zusammenbrechen, einstürzen
colleague	Kollege, Kollegin
to collect	*sammeln, abholen*
collection	*Sammlung*
college	*College, Fach(hoch)schule*
Colombia	Kolumbien
colour	*Farbe*
column	Spalte
to comb; comb	*kämmen; Kamm*
to combine	verbinden, vereinigen
combustion	Verbrennung
combustion chamber	Brennkammer
to come, (came, come)	*kommen*
to come into operation	in Kraft treten
to come true	wahr werden
comfortable	*bequem*
command	Befehl
to comment; comment	kommentieren; Bemerkung, Kommentar
commercial	Werbespot; kaufmännisch
common	gemeinsam, gewöhnlich, verbreitet
common sense	gesunder Menschenverstand
to communicate	*kommunizieren, übermitteln*
communication	*Kommunikation, Übermittlung*
communist era	kommunistisches Zeitalter
community	Bevölkerungsgruppe, Gemeinde
commuter	Pendler
compact	kompakt, komprimiert
company	*Gesellschaft, Firma*
to compare	*vergleichen*
to compete for sth.	um etwas wetteifern, kämpfen
competition	*Wettbewerb, Konkurrenz*
competitive	wettbewerbsfähig
competitor	*Konkurrent(in), Teilnehmer(in)*
to complete; complete	*vervollständigen; vollständig*
complex	komplex, kompliziert
complicated	kompliziert
complication	Komplikation
composting	Kompostierung
compound	(chem.) Verbindung
comprehension	Verständnis
comprehensive school	Gesamtschule
to compress	komprimieren
compression	Kompression, Verdichtung
to concentrate (on)	konzentrieren (auf)
concentration	Konzentration
concept	Konzept, Entwurf
concerned	betroffen, interessiert
concrete	Beton
concurrent	gleichzeitig
condition	Bedingung
conference	Konferenz, Tagung
to confirm	bestätigen
to confront	gegenüberstellen, konfrontieren
confrontation	Konfrontation
congested	überfüllt, verstopft
congestion	Stau
to connect	*verbinden*
connection	*Verbindung, Zusammenhang*
connector	Verbinder, Stecker
conscious	bewusst
consequence	Folge, Konsequenz
consequently	folglich
conservative	konservativ
to conserve	erhalten, schonen
to consider	betrachten, denken (an), berücksichtigen
considerable	beträchtlich
to consist of	bestehen aus
constant	ständig, gleichbleibend
constituency	Wahlkreis
to construct	bauen, konstruieren
construction	Bau, Konstruktion
constructive	konstruktiv
to consult	(sich) beraten, zu Rate ziehen
consumer	Verbraucher(in)
consumption	Verbrauch
to contact; contact	*sich in Verbindung setzen mit; Verbindung, Kontakt*
to contain	*enthalten*
container	*Behälter*
to contaminate	verunreinigen, vergiften
contaminated	verschmutzt, verunreinigt, vergiftet
continent	Kontinent, Erdteil
continental	kontinental
to continue	*fortsetzen, weitermachen, weitergehen*
continuous	ununterbrochen, fortlaufend
contract	Vertrag
contradictory	widersprechend, widersprüchlich
to contrast; contrast	gegenüberstellen, vergleichen; Gegensatz
to contribute	beitragen
contribution	Beitrag
to control; control	kontollieren, steuern; Kontrolle, hier: Regeltechnik
convenient	günstig, praktisch
conventional	herkömmlich, konventionell
conversation	Unterhaltung, Gespräch
to convert	umwandeln
conveyor	Förderband
conveyor belt	Förderband, Fließband
convinced	überzeugt
to cook; cook	*kochen; Koch, Köchin*
cooker	*Herd*
cookie	Keks, Plätzchen
cool box	Kühlfach
to cooperate	*zusammenarbeiten*
cooperation	*Zusammenarbeit*
to cope with	mit etwas zurechtkommen
to copy; copy	*abschreiben, kopieren; Kopie, Exemplar*
corn	Getreide, Mais, Korn
corner	*Ecke*
corporation	Gesellschaft, Körperschaft
to correct; correct	*korrigieren, verbessern; richtig*
correction	*Verbesserung, Korrektur*
to correspond	entsprechen, übereinstimmen, korrespondieren
cosmopolitan	kosmopolitisch
to cost, (cost, cost); cost	*kosten; Kosten*
cost-efficient	kostengünstig
cotton	*Baumwolle*
council	(Stadt-, Grafschafts-) Rat
to count	*zählen, ins Gewicht fallen*
counter	*Ladentisch, Schalter*
counteract	entgegenwirken
country	*Land*
countryside	Landschaft
county	Grafschaft, Bezirk
couple	*(Ehe-)Paar*
courage	Mut
course	*Kurs, Lauf, Gang, Strecke*

to cover; cover	*(be)decken; Abdeckung, Umschlag*
couverage	hier: Ausstattung
cow	*Kuh*
crane	Kran
crankshaft	Kurbelwelle
to crash	zusammenbrechen
crazy	verrückt
to create	schaffen, verursachen
creative	kreativ
critic	Kritiker(in)
critical	kritisch
crops	Feldfrüchte
to cross	*überqueren*
crossing	*Kreuzung*
crossroads	Kreuzung
crowd	*(Menschen-)Menge*
crowded	*dicht gedrängt, voll*
to crush	zerdrücken, zerquetschen
crush zone	Knautschzone
crush-proof	erschütterungssicher
to cry; cry	*schreien, weinen; Schrei*
cultural shock	Kulturschock
culture	Kultur
cup	*Tasse, Pokal*
curious	*neugierig, seltsam*
currency	Währung
current	aktuell, gegenwärtig
curtain	*Vorhang*
customer	*Kunde, Kundin*
customer-orientated	kundenorientiert
customs	*Zoll, Zoll-*
to cut, (cut, cut); cut	*schneiden; Schnitt*
CV = curriculum vitae	Lebenslauf
to cycle	*Fahrrad fahren*
cylinder	Zylinder

D

daffodil	Narzisse
daily	*täglich*
to damage; damage	*beschädigen; Schaden*
damp	feucht
to dance; dance	*tanzen; Tanz*
dancer	Tänzer(in)
danger	*Gefahr*
dangerous	*gefährlich*
dark	*dunkel*
darkness	*Dunkelheit*
dashboard	Armaturenbrett
data	Daten
data processing	Datenverarbeitung
date	*Datum, Zeitpunkt, Verabredung*
date of birth	*Geburtsdatum*
daughter	*Tochter*
day	*Tag*
day by day	Tag für Tag
dead	*tot*
to deal, (dealt, dealt) (in, with); deal	*handeln (mit, von); Geschäft*
dealer	Händler(in)
dear	*lieb*
Dear Madam	Sehr geehrte Dame
Dear Sir	Sehr geehrter Herr
death	*Tod*
to debate; debate	debattieren, diskutieren; Debatte
decade	Jahrzehnt, Dekade, zehn Jahre
decibel	Dezibel (Maß für Lautstärke)
to decide	*(sich) entscheiden*
decision	*Entscheidung*
decision-making process	Entscheidungs-prozess
to decline; decline	sich verschlechtern, sinken; Niedergang, Verschlechterung
to decrease; decrease	abnehmen, vermindern; Abnahme
deep	*tief*
to defend	verteidigen
defender	Verteidiger(in)
definite	eindeutig, bestimmt
degree	akad. Grad, Diplom, Grad (Temperatur)
to delay; delay	verzögern; Verzögerung, Verspätung
to deliver	*liefern, (Rede) halten*
delivery	*Lieferung*
delivery hall	Anlieferungshalle
to demand; demand	verlangen; Nachfrage
democracy	Demokratie
democrat	Demokrat(in)
democratic	demokratisch
to demonstrate	demonstrieren
Denmark	Dänemark
department	*Abteilung*
department store	*Kaufhaus*
to depend (on)	*abhängen (von)*
deposit	Lagerstätte, Guthaben
to describe	*beschreiben*
description	*Beschreibung*
to design; design	*entwerfen, konstruieren; Entwurf, Muster*
designer	Zeichner(in), Designer(in)
desk	*Schreibtisch*
desperate	verzweifelt
destination	Reiseziel, Zielort
to destroy	*zerstören*
destructive	zerstörerisch, destruktiv
destruction	*Zerstörung*
detail	*Einzelheit, Detail*
determination	Entschlossenheit, Bestimmung
to develop	*entwickeln*
developer	Entwickler(in), Bau-unternehmer(in)
developing country	Entwicklungsland
development	*Entwicklung*
device	Gerät
diagram	Diagramm
to dial	*(Telefon) wählen*
dialogue	Dialog
diamond	Diamant
diary	Tagebuch, Terminkalender
to dictate	*diktieren*
dictation	*Diktat*
dictionary	*Wörterbuch*
to die	*sterben*
to differ (from)	*sich unterscheiden (von)*
difference	*Unterschied*
different	*unterschiedlich*
difficult	*schwierig*
difficulty	*Schwierigkeit*
to dig, (dug, dug) (up)	(aus)graben
digital	digital
dining-room	*Esszimmer*
dinner	*(Mittag-, Abend-) Essen*
dinosaur	Dinosaurier
to dip	(in Farbe) eintauchen
direct speech	direkte Rede
to direct (to); direct	*richten, lenken (auf); direkt*
direction	*Richtung*
director	Leiter(in), Direktor(in), Regisseur(in)
dirty	*schmutzig*
disadvantage	*Nachteil*
to disagree (with)	*anderer Meinung sein, nicht über-einstimmen (mit)*
disagreement	abweichende Mei-nung, Meinungs-verschiedenheit
disappearance	Verschwinden
disaster	Katastrophe
discount	Rabatt
to discover	entdecken
to discuss	*besprechen, diskutieren*
discussion	*Gespräch, Unter-redung, Diskussion*
disease	Krankheit
to dislike	*nicht mögen*
to dismantle	zerlegen, demontieren
disorder	Durcheinander, Funktionsstörung
to display; display	zeigen; Anzeige
disposal	Beseitigung, Entsorgung

distance	*Entfernung, Strecke, Abstand*
to distribute	verteilen
distributor	Verteiler
district	Gegend
to disturb	stören
diversion	Umleitung
diversity	Vielfalt, Verschiedenheit
to divide	teilen
DNA sequencing	DNA-Sequenzbildung (-Aufeinanderfolge)
doctor	*Arzt, Ärztin*
document	Dokument, Unterlage
dog	*Hund*
domestic	häuslich
domestic trip	Inlandsausflug
to dominate	beherrschen, dominieren
door	*Tür*
to double	(sich) verdoppeln
double-glazed	doppelt verglast
down	*hinunter, unten*
down payment	Anzahlung
to download	herunterladen
downstairs	*die Treppe hinunter, unten, im Erdgeschoss*
to draft; draft	entwerfen; Entwurf, Konzept
to draw	*zeichnen, ziehen*
draughtsman	Zeichner(in)
drawing	*Zeichnung*
drawing board	Zeichenbrett
to dream; dream	*träumen; Traum*
to dress; dress	*(sich) anziehen; Kleid,Kleidung*
to drink, (drank, drunk); drink	*trinken; Getränk*
to drive, (drove, driven); drive	*fahren, (an)treiben; Fahrt, Laufwerk*
driver	*Fahrer(in)*
to drop	fallen, fallen lassen
drug	Arznei
to dry; dry	*trocknen; trocken*
dual	doppelt, Doppel-
due to	aufgrund von
to dump; dump	(Müll) abladen; Müllkippe
dumping ground	Müllabladeplatz
during	*während*
dustbin	*Mülltonne*
Dutch	niederländisch; Niederländer(in)
duty	*Pflicht, Zoll*
DVD drive	DVD-Laufwerk
DVD recorder	DVD-Rekorder

E

e.g.	z. B. (zum Beispiel)
each	*jede(r,s)*
each other	*einander, sich*
ear	*Ohr*
early	*früh*
to earn	*verdienen*
earnings	Einkommen, Verdienst, Ertrag
earth	*Erde*
earthquake	Erdbeben
east	*Osten, östlich, Ost-, ostwärts*
eastern	Ost-, östlich
easy	*leicht*
easy-to-use	einfach zu bedienen
to eat, (ate, eaten)	*essen*
ecological	ökologisch, umweltfreundlich
ecology	Ökologie
economic	Wirtschafts-, wirtschaftlich
economical	sparsam, wirtschaftlich
economics	Wirtschaft(swissenschaft)
economist	Wirtschaftswissenschaftler(in)
economy	*Wirtschaft*
edge	*Kante, Rand, Schneide*
edition	Ausgabe
editor	Herausgeber(in), Redakteur(in)
to educate	*erziehen*
education	*Erziehung*
to effect; effect	*durchführen, erzielen, leisten; Wirkung*
efficiency	Leistungsfähigkeit
efficient	leistungsfähig
effort	Anstrengung, Mühe, Versuch
egg	*Ei*
either ... or	*entweder ... oder*
to eject	auswerfen
to elect	*wählen*
election	*Wahl*
electric	*elektrisch*
electrical engineering	Elektrotechnik
electrically-driven	elektrisch angetrieben
electrician	Elektriker(in)
electricity	*Elektrizität*
Electricity Board	Elektrizitätsversorgungsunternehmen
electronic	elektronisch
electronics	*Elektronik*
elevator	Aufzug, Lift
to eliminate	beseitigen
else	*sonst*
elsewhere	woanders
embassy	Botschaft
emergency	*Notfall*
to emigrate	auswandern
emigration	Auswanderung
emission	Emission, Ausstoß
emphasis	Betonung
to emphasize	betonen
to employ	*beschäftigen, einstellen*
employed	*beschäftigt*
employee	*Angestellte(r), Arbeitnehmer(in)*
employer	*Arbeitgeber(in)*
employment	*Beschäftigung, Anstellung, Arbeit*
empty	*leer*
to enable (s.o. to do sth.)	es (jmdm.) möglich machen (etwas zu tun)
to enclose	beilegen
to end; end	*(be)enden; Ende*
to endanger	gefährden
energy	*Energie, Kraft*
energy recovery	Energierückgewinnung
engine	*Motor, Triebwerk, Lokomotive*
engineer	*Ingenieur(in), Techniker(in)*
engineering	*Technik; technisch*
engineering studies	etwa: Fachbereich Technik
English Channel	Ärmelkanal
to enjoy	*genießen, sich freuen an*
enjoyable	*schön, angenehm, unterhaltsam*
enormous	gewaltig, enorm
enough	*genug*
to enquire (about)	*sich erkundigen (nach), fragen (nach)*
enquiry	*Anfrage, Erkundigung, Untersuchung*
to enter	*eintreten*
enthusiasm	Begeisterung, Enthusiasmus
enthusiast	Begeisterte(r), Enthusiast(in)
enthusiastic	begeistert
entire	ganz, komplett
entrance	*Eintritt, Eingang*
entrance exam	Aufnahmeprüfung
entrepreneurship	Unternehmertum, unternehmerische Vorhaben
entry	Einreise, Eingang, Eintritt
envelope	*Umschlag*
environment	*Umwelt*
environmental	Umwelt-
environmentalist	Umweltschützer(in)
equal	*gleich*
to equip	ausrüsten
equipment	*Ausrüstung, Ausstattung*
to erect	errichten, aufbauen
especial	*besondere(r,s)*
essential	wesentlich
to establish	schaffen, gründen, errichten
to estimate	schätzen
estimated value	geschätzter Wert

Estonia	Estland
etc. (et cetera)	und so weiter
European Commission	Europäische Kommission
European Union	Europäische Union
even	*sogar, selbst, eben, gleich*
evening	*Abend*
event	*Ereignis*
eventual(ly)	am Ende, schließlich
ever	*je*
every	*jede(r,s)*
everybody	*jeder*
everyone	*jeder*
everything	*alles*
everywhere	*überall*
evolution	Entwicklung, Evolution
to evolve	(sich) entwickeln
exact	*genau*
exaggeration	Übertreibung
exam(ination)	*Prüfung*
to examine	untersuchen, überprüfen
example	*Beispiel*
excellent	*ausgezeichnet, hervorragend*
except	außer
exchange	Austausch
exchange rate	Wechselkurs
excitement	Aufregung, Begeisterung
exciting	*aufregend*
to excuse; excuse	*(sich) entschuldigen; Entschuldigung*
exercise	*Übung*
exhaust pipe	Auspuffrohr
to exist	*existieren, bestehen*
existence	Existenz, Dasein
exit	*Ausgang, Ausfahrt*
to expand	ausdehnen
expansion	Ausweitung
to expect	*erwarten*
expenditure	Ausgabe(n), Aufwand
expensive	*teuer*
to experience	erleben, erfahren
experience	*Erfahrung*
expert	Experte, Expertin
to explain	*erklären*
explanation	*Erklärung*
to export; export	*ausführen, exportieren; Ausfuhr, Export*
export assistant	etwa: Außenhandelskaufmann/-frau
exposed	ungeschützt, ausgesetzt
to express; express	*ausdrücken, äußern; Schnellzug; als Eilsache*
expression	*Ausdruck*
extension	Streckung, Verlängerung, Erweiterung, Durchwahl (Telefon)
to a great extent	in großem Umfang
extinction	Aussterben
extra	*besonders*
to extract; extract	gewinnen, entnehmen; Auszug, Extrakt
extreme	extrem, äußerst, höchst
eye	*Auge*

F

to face; face	*sich gegenübersehen, gegenübertreten; Gesicht*
facility	Einrichtung, Möglichkeit
fact	*Tatsache*
factor	Faktor
factory	*Fabrik*
to fade (away)	verblassen, nachlassen, verwelken
to fail	scheitern
fair	*Messe, Markt; gerecht, fair*
fairytale	Märchen
to fall asleep	einschlafen
to fall, (fell, fallen); fall	*fallen; Fallen, Sturz, Herbst (AE)*
false	*falsch*
family	*Familie*
famous	*berühmt*
fanatic	fanatisch; Fanatiker(in)
far	*weit*
fare	*Fahrpreis*
farm	*Bauernhof, Farm*
farmer	*Landwirt(in), Bauer, Bäuerin*
farming	*Landwirtschaft*
fascination	Faszination
fashionable	modisch
fast	*schnell*
fast running	schnell fließend
fat	*Fett; dick*
father	*Vater*
fault	Fehler, Schuld, Mangel
favourite	*Lieblings-; Liebling*
fear	Furcht
feature	Merkmal, Bestandteil
to feed, (fed, fed)	*füttern*
feedback	Rückwirkung, Feedback
to feel, (felt, felt)	*fühlen*
feeling	*Gefühl*
female	weiblich
fertilizer	Dünger
few	*wenige*
field	*Feld*
to fight, (fought, fought); fight	*kämpfen; Kampf*
figure	Zahl, Figur
file	Akte, Daten
to fill	*füllen*
final	*Finale, Endrunde; letzte(r,s), endgültig*
finance	Finanzwesen
financial	finanziell
financial services	Finanzdienstleistungen
to find, (found, found)	*finden*
fine	*Strafe; fein, gut, schön*
to finish; finish	*beenden; Ziel, Vollendung*
to fire; fire	*feuern; Feuer*
fire-proof	feuersicher, brandsicher
firm	*Firma; fest, verbindlich*
first	*erste(r,s), zuerst*
first of all	*zuerst, vor allem*
to fish; fish	*fischen; Fisch*
to fit; fit	*(zusammen)passen, einbauen; gesund, in Form; geeignet*
to fix	*befestigen, reparieren, besorgen*
fixed-route	schienengebunden
flag	Fahne
flair	Flair, Gespür, Talent, Begabung
flame	Flamme
flat	*Wohnung; flach*
to flee (fled, fled)	fliehen, flüchten
flexibility	Flexibilität
flexible	flexibel
flight	*Flug*
flight attendant	Flugbegleiter(in)
to flood; flood	überfluten; Flut
floor	*Boden*
flower	*Blume*
to fly, (flew,flown); fly	*fliegen; Fliege*
flyover	Überführung
foam	Schaum
fog	Nebel
to follow	*folgen*
food	*Nahrung, Essen*
food stamp	Essensmarke
foot	*Fuß*
football	*Fußball*
footballer	Fußballspieler(in)
for better or for worse	zum Besseren oder Schlechteren hin
for example	*zum Beispiel*
for instance	beispielsweise
force	Kraft, Stärke, Gewalt, Macht
foreign	*ausländisch, fremd*
foreign	ausländisch
forerunner	Vorläufer
forest	Wald
forever	*für immer*
to forget, (forgot, forgotten)	*vergessen*
fork	*Gabel*

to form; form	*gestalten; Form, Formular, Klasse*
former	ehemalig
formula	Formel
fortunate	*glücklich*
fortune	Glück, Reichtum
forward(s)	*vorwärts*
fossil fuels	fossile Brennstoffe
to found	gründen
foundation	Gründung, Fundament
four-stroke engine	Viertaktmotor
frame	Rahmen
France	Frankreich
to free; free	*befreien; frei*
freedom	Freiheit
freeway	*Autobahn (AE)*
French	französisch; Franzose, Französin
frequent	*häufig*
fresh	*frisch*
freshwater	Süßwasser
fridge	*Kühlschrank (Kurzform)*
fried	gebraten
friend	*Freund(in)*
friendly	*freundlich*
Friends of the Earth	Name einer Umweltschutzorganisation
to frighten	erschrecken, Angst machen
front	*Vorderseite*
front-page	Titelseite
fruit	*Frucht, Früchte, Obst*
fuel	Kraftstoff, Brennstoff
fuel cell	Brennstoffzelle
to fulfil	erfüllen, entsprechen
full	*voll*
full-time	Vollzeit
fumes	Rauch, Dämpfe, Abgase
fun	*Spaß*
function	Funktion, Aufgabe
fund	Fond, Fundus, Mittel
furious	wütend, wild, heftig
furnace	Ofen
furniture	*Möbel*
further	*weiter*
furthermore	außerdem
future	*Zukunft; zukünftig*

G

gadget	Gerät, Apparat
to gain; gain	gewinnen; Gewinn
galactic	galaktisch, Weltraum-
gallon	Gallone
game	*Spiel*
garage	*Autowerkstatt, Garage*
garbage	Abfall, Müll
garden	*Garten*
gate	*Tor*
to gather	sich versammeln
GCSE = General Certificate of Secondary Education	etwa: Fachoberschulreife
gear	Getriebe, Gang
gearbox	Getriebe
general	allgemein
general science	allgemeine Naturwissenschaften
to generate	erzeugen
generator	Generator, Lichtmaschine
genetic	genetisch, Erb-
genetic research	Genforschung
genetically modified	gentechnisch verändert
gentleman	*Herr*
geo-thermal	geothermisch (Erdwärme nutzend)
geographical	geographisch
Germany	Deutschland
to get, (got, got)	*bekommen, werden*
to get married	*heiraten*
to get on like a house on fire	sich sehr gut verstehen
to get on with	auskommen mit
to get rid of	loswerden, abschaffen
to get to know	kennenlernen
to get up	*aufstehen*
to get used to	sich gewöhnen an
girl	*Mädchen*
girlfriend	*Freundin*
to give, (gave, given)	*geben*
to give up	*aufgeben*
glad	*froh*
glance	Blick
glass	*Glas*
glasses	*Brille*
glass-walled	mit Wänden aus Glas
global	global, weltweit
global warming	weltweiter Temperaturanstieg
glory	Ruhm
glove	*Handschuh*
to go down	*untergehen, sinken, fallen*
to go on	*weitergehen, weitermachen*
to go on strike	in den Streik treten
to go shopping	*einkaufen*
to go up	*hinaufgehen, wachsen, steigen*
to go, (went, gone)	*gehen*
goal	Ziel, Tor
goat	Ziege
good	*gut*
goodbye	*auf Wiedersehen*
goods	*Güter, Waren*
to govern	*regieren*
government	*Regierung*
GPS (Global Positioning System)	GPS-Navigationssystem
gradual	allmählich, sanft (ansteigen, abfallen)
to graduate	das Studium beenden, einen akad. Grad erwerben
grandfather	*Großvater*
grandmother	*Großmutter*
grandparents	*Großeltern*
grant	Zuschuss, Bafög
graph	Schaubild, mathematische Kurve
grateful	*dankbar*
great	*großartig, groß*
Great Britain	Großbritannien
Greece	Griechenland
Greek	griechisch; Grieche, Griechin
green	*grün*
greenhouse	Treibhaus, Gewächshaus
greenhouse effect	Treibhauseffekt, globale Erwärmung
to greet	*grüßen*
grey	*grau*
grid	Versorgungsnetz
to grip; grip	Griff, Halt; (er)greifen, fesseln
grocer	*Lebensmittelhändler(in)*
ground	*Boden*
group	*Gruppe*
to grow, (grew, grown)	*wachsen*, anbauen
to grow up	aufwachsen
grown-up	*Erwachsene(r); erwachsen*
growth	*Wachstum*
to guess	*erraten*
guest	*Gast*
to guide	leiten, führen
guideline	Richtlinie
guilty	schuldig

H

habit	Gewohnheit
hair	*Haar*
hairdresser	Friseur(in)
half	*Hälfte; halb*
hall	*Flur, Saal*
to halt; halt	anhalten; Halt, Pause, Unterbrechung
ham	*Schinken*
handbag	*Handtasche*
handbook	Handbuch
handful	Handvoll
handkerchief	*Taschentuch*
to hang, (hung, hung)	*(auf)hängen*
to happen	*geschehen*
happiness	Glück, Heiterkeit, Zufriedenheit

happy	*glücklich, heiter*
harbour	*Hafen*
hard	*hart, schwierig, anstrengend*
hard-working	fleißig
hardly	*kaum*
to harm	schädigen
harmful	schädlich
hat	*Hut*
to hate	*hassen*
to have, (had, had)	*haben, lassen*
to have in common	gemeinsam haben
to have to	*müssen*
hazard	Gefahr
to head; head	*anführen, fahren nach; Kopf, Leiter*
head of department	hier: Fachbereichsleiter
to head off	(hin) fahren (nach), losfahren
headache	*Kopfschmerz*
heading	Überschrift
headlight	Scheinwerfer
headline	*Überschrift*
headset	Kopfhörer
health	*Gesundheit*
health care	Gesundheitsfürsorge
healthy	*gesund*
to hear, (heard, heard)	*hören*
heart	*Herz*
heart attack	Herzanfall
to heat; heat	*heizen; Hitze*
heating	*Heizung*
heavy	*schwer*
heavy industry	Schwerindustrie
heavy metal	Heavy Metal (harte Rockmusik)
height	*Höhe*
hello	*Hallo*
helmet	Helm
to help; help	*helfen; Hilfe*
helpful	*hilfreich*
here	*hier*
here you are	*bitte(schön)!*
hereditary peer	Peer mit vererbtem Adelstitel
to hesitate	*zögern*
Hi Tech (High Technology)	*Hochtechnologie*
to hide, (hid, hidden)	*(sich) verstecken*
high	*hoch*
high class	hochwertig, erstklassig
high-speed	Hochgeschwindigkeits-
highway	Landstraße
hill	*Hügel*
historical	historisch
history	*Geschichte*
to hit, (hit, hit); hit	*schlagen; Schlag, Treffer, Erfolg*
to hold, (held, held)	*halten*

hole	*Loch*
holiday	*freier Tag, Feiertag, Urlaub*
holidaymaker	Urlauber(in)
Holy Bible	Heilige Bibel
home	*Wohnung, Heimat; zu/nach Hause*
homework	*Hausaufgabe*
honest	*ehrlich*
to hope; hope	*hoffen; Hoffnung*
hopeful	zuversichtlich, vielversprechend
horrified	entsetzt
horse	*Pferd*
hospital	*Krankenhaus*
host	Gastgeber(in)
hot	*heiß*
hour	*Stunde*
hourly	stündlich
house	*Haus*
House of Commons	Unterhaus
House of Lords	Oberhaus
household	Haushalt
housewife	*Hausfrau*
housework	*Hausarbeit*
housing	Unterkunft, Wohnungen
housing estate	Siedlung
how	*wie*
how are you?	*wie geht es Dir/ Ihnen?*
however	*jedoch*
huge	riesig
human	menschlich
humanity	Menschheit, Menschlichkeit
humanoid	menschenartig
hundred	*Hundert*
Hungary	Ungarn
hungry	*hungrig*
to hurry; hurry	*sich beeilen; Eile*
to hurt	*verletzen*
husband	*Ehemann*
hybrid	Hybrid-, Misch-; Kreuzung, Mischung
hydro	Wasser-
hydro-electric	hydroelektrisch (Wasserkraft nutzend)
hydrogen	Wasserstoff
hydrogen-based	auf Wasserstoffbasis arbeitend
hypnotic	hypnotisch

I

ice	Eis
ice cream	*Eis*
icon	Idol, Ikone
idea	*Idee*
to identify	identifizieren
identity	Identität, Übereinstimmung
if	*wenn, falls*
to ignite	zünden
ignition	Zündung

to ignore	ignorieren, nicht beachten
ill	*krank*
illegal	illegal, gesetzlich unzulässig
illness	*Krankheit*
to imagine	*sich vorstellen*
immediate	*unmittelbar, unverzüglich*
immigrant	*Einwanderer, Einwanderin*
to immigrate	*einwandern*
immigration	Einwanderung
immune system	Immunsystem
to implant	einpflanzen
to implement	erfüllen, umsetzen, ausführen
to implode	implodieren
to import; import	*importieren, einführen; Einfuhr, Import*
importance	*Wichtigkeit*
important	*wichtig*
impossible	*unmöglich*
to impress	beeindrucken
impression	Eindruck
impressive	beeindruckend
to improve	*(sich) verbessern*
improvement	Verbesserung
in addition	zusätzlich
in case of	im Falle von
in charge of	verantwortlich für
in conclusion	abschließend
in contrast	im Gegensatz
in fact	*tatsächlich*
in favour of	zugunsten von
in front of	*vor*
in general	im Allgemeinen
in italics	*kursiv*
in my opinion	*meiner Meinung nach*
in my view	meiner Ansicht nach
in order to	um zu
(in) preference (to)	(unter) Bevorzugung (von)
in the end	schließlich
in the first place	zuerst
in the long run	auf lange Sicht
(in the) meantime	*inzwischen*
in the mid	in der Mitte von, in der Mitte der …
in time	*rechtzeitig*
inch	*(Maßeinheit) Zoll*
incineration	Verbrennung
to be inclined (to do sth.)	dazu neigen, etwas zu tun
to include	*einschließen*
income	*Einkommen*
to increase	*zunehmen, steigen, erhöhen*
incredible	unglaublich
indeed	*tatsächlich*
independent	unabhängig
India	Indien
individual	individuell

to induce	verursachen, herbeiführen
industrial	industriell
industrialized	industrialisiert
industry	*Industrie*
infertile	unfruchtbar
infinitive	Infinitiv, Grundform
inflexible	unflexibel
inflow	Zustrom
to influence	beeinflussen
to inform	*informieren*
informal	informell
information sheet	Informationsblatt
to inject	injizieren, spritzen
injury	Verletzung
ink	Tinte
ink jet printer	Tintenstrahldrucker
inner	Innen-, inner-
innocence	Unschuld
innovation	Neuerung
innovative	innovativ
to insert	einlegen (Diskette, Kassette)
inside	*in, innerhalb*
to install	einbauen
installation	Installation, Anschluss
instead (of)	*anstatt*
institute	Institut
to instruct	*unterrichten, anweisen*
instruction	*Unterricht, Anweisung, (Gebrauchs-) Anleitung*
instrumentation	Messtechnik
to insulate	isolieren
insulation	Isolierung
insurance	*Versicherung*
to insure	versichern
intake	Aufnahme, Ansaugen
to integrate	integrieren
to intend	beabsichtigen
interconnected	vernetzt
interconnectivity	Vernetzung
intercontinental	interkontinental
interest	*Interesse, (Plural) Zinsen*
to be interested (in)	interessiert sein (an)
interesting	*interessant*
interface	Schnittstelle
internal combustion engine	Verbrennungsmotor
interval	Abstand
to be into sth.	Fan von etwas sein
to introduce	*einführen, vorstellen*
introduction	*Einführung, Vorstellung*
introductory	einleitend
to invent	*erfinden*
invention	*Erfindung*
inventor	Erfinder(in)
to invest	*investieren*
investment	Investition, (Kapital-) Anlage
invitation	*Einladung*
to invite	*einladen*
to involve	beteiligen, verwickeln
Ireland	Irland
irreplaceable	unersetzlich
irrigation	Bewässerung
island	*Insel*
to isolate	isolieren
issue	Frage, Angelegenheit, Problem
Italy	Italien
itself	selbst

J

jacket	*Jacke*
jam	*Marmelade*
Japanese	japanisch; Japaner(in)
jersey	Trikot
job	*Arbeit, Stelle, Aufgabe*
jobless	arbeitslos
to join	*(sich) anschließen, beitreten*
joint	gemeinsam
journey	*Reise*
juice	*Saft*
to jump; jump	*springen; Sprung*
just	*genau, soeben, nur*

K

to keep, (kept, kept)	*(be)halten*
key	*Schlüssel*
key point	springender Punkt
to kill	*töten*
kilogram(me)	*Kilogramm*
kind	*Art, Sorte; freundlich*
king	*König*
to kiss; kiss	*küssen; Kuss*
kit	Ausrüstung, Bausatz, Set
kitchen	*Küche*
knee	*Knie*
knee-cap	Kniescheibe
knife	*Messer*
to knock; knock	*klopfen; Klopfen*
to know, (knew, known)	*wissen*
knowledge	*Wissen*
kph	km/h (Kilometer pro Stunde)

L

lab	Labor
labour	Arbeit
labour market	Arbeitsmarkt
lack	Mangel
ladder	*Leiter*
lady	*Dame*
lake	*See*
lamp	*Lampe*
land	*Land*
landfill (deposit)	Mülldeponie
landlady	*Vermieterin, Wirtin*
landlord	*Vermieter, Wirt*
landscape	Landschaft
lane	(Fahr-)spur, Landstraße, Weg
language	*Sprache*
large	*groß*
to last; last	*dauern; letzte(r,s)*
late	*spät*
later on	*später*
latest	*neueste(r,s)*
Latin America	Lateinamerika
to laugh	*lachen*
laughter	*Gelächter*
law	*Gesetz*
to lay, (laid, laid)	*legen*
to lay the table	den Tisch decken
lazy	*faul*
lead	Leitungskabel, Zuleitung
to lead, (led, led)	*führen, leiten*
leadership	Führung, Leitung
leaflet	*(Hand-)Zettel, Flugblatt*
league	Liga
lean management	„schlankes" Management
to learn, (learnt, learnt)	*lernen*
learner	*Lernende(r), Anfänger(in)*
least	*wenigste(r,s), geringste(r,s)*
to leave, (left, left)	*verlassen*
lecture	Vorlesung
left	*links*
(to feel) left out	(sich) ausgeschlossen (fühlen)
leg	*Bein*
legal	legal, gesetzlich zulässig
legal services	Rechtsberatungen, -dienstleistungen
legislation	Gesetzgebung
leisure	*Freizeit*
lemon	*Zitrone*
to lend, (lent, lent)	*verleihen*
length	*Länge*
less	*weniger*
lesson	*Unterrichtsstunde*
to let, (let, let)	*lassen*
letter	*Brief*
level	Stufe, Niveau
library	Bibliothek
to lie, (lay, lain)	*liegen*
to lie, (lied, lied)	*lügen*
life	*Leben*
life peers	Peer auf Lebenszeit
lifestyle	Lebensstil
lifeless	leblos

to lift; lift	*anheben,aufheben; Aufzug, Mitfahrgelegenheit*
light	*Licht, Lampe; hell, leicht*
light industry	Leichtindustrie
to light, (lit, lit)	anzünden, (be)leuchten
lighting	Beleuchtung
to like; like	*mögen; wie*
likely	wahrscheinlich
limit	*Grenze, Beschränkung*
line	*Linie, Zeile*
lined with	gesäumt von
to link; link	verbinden; Verbindung
to link together	miteinander verbinden
lip	*Lippe*
to list; list	*aufführen, auflisten; Liste*
to listen	*zuhören*
listener	Zuhörer(in)
literature	Literatur, Informationsmaterial
litre	Liter
litter	Streu, Stroh
little	*klein, wenig*
to live; live	*leben, wohnen; lebend, direkt, live*
to live out (dream)	(Traum) verwirklichen
living room	*Wohnzimmer*
to load; load	*laden; Ladung*
loaded with	überladen mit
local	*lokal, ortsansässig, örtlich*
Local Education Authority (L.E.A.)	örtliche Schulbehörde
locals	Einwohner
location	Lage, Standort
logical	logisch
logo	Logo, (Firmen)Emblem
long	*lang*
long-term	langfristig
long time no see	lange nicht gesehen
to look	*schauen, (aus)sehen*
to look after	*sich kümmern um*
to look for	*suchen*
to look forward to	*sich freuen auf*
to look into	untersuchen, prüfen
to look like	aussehen wie
to look out	*aufpassen, hinaussehen*
to look up	*nachschlagen, heraussuchen, aufblicken*
to loot	plündern
lorry	*Lastwagen*
to lose, (lost, lost)	*verlieren*
loss	*Verlust*
(a) lot	*viel; Los*

lots of	*viel*
loud	*laut*
to love; love	*lieben; Liebe*
lovely	*hübsch*
low	*niedrig*
low-cost	preiswert
to lower	senken, herabsetzen
luck	*Glück, Schicksal*
lucky	*glücklich (im Sinne von: Glück haben)*
luggage	*Gepäck*
lunch	*Mittagessen*
Luxembourg	Luxemburg

M

machine	*Maschine*
machinery	Maschinen
mad	*verrückt*
madam	*Gnädige Frau*
magazine	*Zeitschrift*
mail	*Post*
main	*Haupt-*
mainly	*hauptsächlich*
to maintain	behaupten
maintenance	Wartung, Instandhaltung
major	bedeutend
major road	Hauptstraße
majority	Mehrheit
to make, (made, made)	*machen, herstellen*
to make matters worse	zu allem Unglück
to make sense	einen Sinn ergeben
to make sure	sich überzeugen
to make up one's mind	sich entschließen
maker	Hersteller(in)
Malaysia	Malaysien
male	männlich
mammal	Säugetier
man	*Mann*
to manage	*zu Stande bringen, schaffen, leiten, verwalten*
management	*(Geschäfts-)Leitung, Verwaltung, Durchführung*
managing director	Geschäftsführer(in)
mangrove swamp	Mangrovensumpf
to manipulate	manipulieren
mankind	Menschheit
manpower	Arbeitskräfte
to manufacture	herstellen
manufacturer	Hersteller(in)
manure	Dung
many	*viele*
map	*(Land-)Karte*
marital status	Familienstand
mark	Zensur, Note
market	*Markt*
market development	Marktentwicklung

marmalade	*(Orangen-) Marmelade*
marriage	Ehe, Hochzeit
married	*verheiratet*
to marry	*heiraten*
mashed	zerdrückt
mashed potatoes	Kartoffelbrei
mass	*Masse*
massive	massiv, wuchtig, enorm
to match; match	*zusammenbringen, passen; Wettkampf, Streichholz*
mathematics	*Mathematik*
to matter; matter	*etwas ausmachen; Angelegenheit*
may	*dürfen*
maybe	*vielleicht*
mayor	*Bürgermeister(in)*
meal	*Mahlzeit*
to mean, (meant, meant)	*bedeuten, meinen*
meaning	*Bedeutung*
means	Mittel, Möglichkeiten
meanwhile	*inzwischen*
measure	Maßnahme
meat	*Fleisch*
mechanic	*Mechaniker(in)*
mechanical engineering	Maschinenbau
mechanism	Mechanismus
media	*Medien*
medical	ärztlich, medizinisch
medicine	Medizin, Medikament
medicine cabinet	Arzneimittelschrank
Mediterranean (Sea)	Mittelmeer
medium-sized	mittelgroß
to meet, (met, met)	*(sich) treffen, kennenlernen*
to meet the growing demand	der wachsenden Nachfrage gerecht werden
meeting	*Treffen*
megawatt	Megawatt
to melt	schmelzen
member	*Mitglied*
membership	Mitgliedschaft
membrane	Membrane
memory	Gedächtnis
to mend	*reparieren*
to mention	*erwähnen*
menu	Speisekarte
message	*Mitteilung, Nachricht*
metal	Metall
methane	Methan(-gas)
method	*Methode*
metre	*Meter*
Mexican	Mexikaner(in); mexikanisch
Mexico	Mexiko
microphone	Mikrofon
microprocessor	Mikroprozessor

microwave	Mikrowelle
middle	Mitte
midnight	Mitternacht
might	Gewalt, Macht; könnte(n)
migration	(Ab-)Wanderung
mile	Meile
mileage	Meilen, Meilenstand
milk	Milch
to milk	melken
millennium	Jahrtausend
Millennium Wheel	Millennium-Riesenrad (in London)
to mind; mind	etwas ausmachen; Meinung, Gedanken, Verstand
mine	Bergwerk
mining	Bergbau
minor	kleiner, weniger bedeutend, leicht
minority	Minderheit
minus	minus, ohne, abzüglich
minute	Minute
mirror	Spiegel
Miss	Fräulein
to miss	vermissen, verfehlen, verpassen
mistake	Fehler
misty	neblig, dunstig
to mix; mix	mischen; Mischung
to mix up	vertauschen
mixture	Mischung
mobile phone	Handy
mobility	Mobilität
mobilization	Mobilisierung
modal	Hilfsverb
model	Modell
to modify	verändern, modifizieren
molecular	molekular
molecular bond	molekulare Verbindung
moment	Augenblick
money	Geld
to monitor	überwachen
monotonous	eintönig, monoton
month	Monat
moon	Mond
moor	Moor
more	mehr
moreover	überdies, außerdem
morning	Morgen
most	die meisten
mother	Mutter
motorist	Autofahrer(in)
motorway	Autobahn
mountain	Berg
mouth	Mund
to move; move	bewegen, umziehen; Umzug
movement	Bewegung
MP (Member of Parliament)	Mitglied des Parlaments
mph (miles per hour)	Meilen pro Stunde
Mr	Herr
Mrs	Frau (verheiratet)
Ms	Frau (verheiratet oder unverheiratet)
much	viel
multi-	Mehr-, Viel-
multi-storey car park	Parkhaus
music	Musik
must	müssen
must not	nicht dürfen

N

to name; name	benennen; Name
namely	nämlich
narrow	eng
NASA (National Aeronautics and Space Administration)	Raumfahrtbehörde
nationality	Nationalität
nationwide	landesweit
native-born	gebürtig
natural	natürlich
nature	Natur
navigation system	Navigationssystem
near	nahe
nearby	nahe gelegen
nearly	fast, beinahe
neat	ordentlich, gelungen, schlau
necessary	notwendig
to need; need	benötigen; Notwendigkeit
negative	negativ
to neglect	vernachlässigen
neighbour	Nachbar(in)
neither … nor	weder … noch
net	Netz
Netherlands	Niederlande
network	Netz
never	nie
nevertheless	dennoch
new	neu
news	Nachrichten, Neuigkeit
newspaper	Zeitung
next	nächste(r,s)
nice	nett
night	Nacht
no	kein; nein
no longer	nicht mehr
no one	niemand
nobody	niemand
noise	Lärm
noisy	geräuschvoll
non-	Nicht-
non-returnable	Einweg-
non-threatening	harmlos, nicht bedrohlich
noon	Mittag
north	Norden; nördlich, Nord-, nordwärts-
nose	Nase
not	nicht
not any more	nicht länger
not either	auch nicht
not even	nicht einmal
not yet	noch nicht
to note; note	notieren; Notiz
nothing	nichts
to notice; notice	bemerken; Aushang, Kenntnis, Beachtung
noun	Substantiv
novel	Roman
now	jetzt
nowadays	heutzutage
nowhere	nirgendwo
nuclear	nuklear, Kern-, Atom-
number	Zahl, Nummer, Anzahl
numerical	numerisch, zahlenmäßig
nurse	Krankenschwester

O

o'clock	Uhr (Zeitangabe)
object	Gegenstand
objective	Ziel
to obtain	bekommen, erhalten
obvious	offensichtlich
to occupy	einnehmen, besetzen
to occur	auftreten
ocean	Meer, Ozean
of course	natürlich
off	von … weg
off-shore	Off-shore, im Meer
off-the-grid	nicht an das Versorgungsnetz angeschlossen
to offer; offer	anbieten; Angebot
office	Büro, Amt
office skills	Bürofertigkeiten
office worker	Büroangestellte(r)
officer	Beamte(r), Offizier(in)
official	offiziell; Beamte(r), Funktionär(in)
often	oft
oil	Öl
old	alt
(on the) left-hand (side)	auf der linken (Seite)
on the one hand	auf der einen Seite
on the other hand	auf der anderen Seite
on the right	rechts
on the road	auf dem Weg, unterwegs
on time	pünktlich
on top of	zusätzlich zu
on your own	allein
once	einmal
once again	noch einmal
only	nur
to open; open	öffnen; offen
to operate	bedienen, verkehren
operator	Bediener(in), Telefonvermittler(in)
opinion	Meinung
opportunity	Gelegenheit, Möglichkeit
opposite	Gegensatz; gegensätzlich

optical	optisch
or	*oder*
to order; order	*bestellen, befehlen; Bestellung, Befehl*
ordinary	gewöhnlich
ore	Erz
organization	*Organisation*
to organize	*organisieren*
organizer	Veranstalter(in)
original	ursprünglich, originell
other	*andere(r,s)*
otherwise	sonst, anderenfalls
ought to	sollte(n)
out of town	außerhalb der Stadt
out of work	*arbeitslos*
out-of-date	veraltet, überholt, altmodisch
outbound	ins Ausland
to outdo	übertreffen, überbieten
outdoor	draußen, im Freien
output	(Produktions-) Leistung
outside	*Außenseite; draußen, außerhalb*
outsider	Außenseiter(in)
outstanding	hervorragend, außergewöhnlich
over and over again	immer wieder
over-training	zu häufiges Training
over-use	zu häufiger Gebrauch
overcrowded	überfüllt
to overestimate	überschätzen
overseas	in Übersee
to oversee	überwachen, leiten
to own; own	*besitzen; eigene(r,s)*
owner	*Eigentümer(in)*
ozone hole	Ozonloch
ozone layer	Ozonschicht

P

pace	Tempo, Schritt
to pack; pack	*packen; Packung*
package tour	Pauschalreise
packaging	Verpackung
packet	*Paket*
packing	Verpackung
page	*Seite*
pain	Schmerz
to paint; paint	*malen; Farbe*
pair	*Paar*
panel	Schalttafel
paper	*Papier*
paradise	Paradies
paragraph	Absatz
to paraphrase; paraphrase	umschreiben; Umschreibung
parents	*Eltern*
to park; park	*parken; Park*
parliament	Parlament
parliamentary	parlamentarisch
part	*Teil*
part-time	Teilzeit
to participate	teilnehmen
particular	besondere(r,s)
to pass	*vorbeigehen, (Prüfung) bestehen*
to pass away	entschlafen, hinscheiden, sterben (Umschreibung)
passenger	*Passagier, Fahrgast*
passive (voice)	Passiv
passport	*Pass*
past	*Vergangenheit*
past tense	Vergangenheitsform
patience	Geduld
pavement	Bürgersteig (BE), Straße (AE)
to pay attention	aufmerksam sein
to pay, (paid, paid); pay	*(be)zahlen; Bezahlung*
payment	Zahlung
peace	*Friede*
peaceful	friedlich
to peak; peak	seine Spitzenform erreichen; Spitze, Gipfel, Höhepunkt
peak time	Stoßzeit
peat	Torf
peat bog	Torfmoor
pedestrian	Fußgänger(in)
pedestrian mall	Fußgängerzone
pen	*Füllhalter*
pence	*Plural von Penny*
pencil	*Bleistift*
penny	*Penny (engl. Münze)*
people	*Leute; Volk*
per	pro
per head/capita	pro Kopf
per cent	*Prozent*
percentage	prozentualer Anteil
perfect	*vollkommen, perfekt*
to perform	ausführen, vollbringen
perhaps	*vielleicht*
period	*Periode, Zeitraum, Schulstunde*
permanent	(be)ständig, bleibend
permission	Erlaubnis
persecution	Verfolgung
personal	*persönlich*
personality	Persönlichkeit
personnel	Personal
pesticide	Pestizid
petrol	*Benzin*
petrol station	Tankstelle
pharmaceutical	pharmazeutisch
pharming	(Kunstwort) Landwirtschaft betreiben zu pharmazeutischen Zwecken
to phone; phone	*telefonieren; Telefon*
phone call	Anruf
photograph	*Foto*
photographer	Fotograf(in)
photovoltaic (pv)	photoelektrisch
phrase	Ausdruck
physical	physikalisch, physisch
to pick	*pflücken, wählen*
to pick up	*nehmen, aufheben*
picture	*Bild*
piece	*Stück*
pile	Pfeiler, Pfahl, Haufen
pill	Pille
pipe	*Pfeife, Röhre*
to pinpoint	genau anzeigen, genau bestimmen
piston	Kolben
piston rod	Pleuel-/Kolbenstange
pit	Grube
pitch	Spielfeld
to place; place	*stellen, legen; Platz*
to plan; plan	*planen; Plan*
plane	*Flugzeug*
planner	Planer(in)
to plant; plant	*pflanzen; Pflanze, Werk*
plastic	Kunststoff, Plastik
plate	*Teller*
platform	*Bahnsteig*
to play; play	*spielen; (Schau-) Spiel*
player	*Spieler(in)*
playground	*Spielplatz*
pleasant	angenehm
please	*bitte*
pleased	zufrieden, erfreut
pleasure	*Vergnügen*
plentiful	reichlich, häufig
plenty	viel, eine Menge
to plot	zeichnen, abstecken
to plug	stecken
pocket	*Tasche*
pocket money	Taschengeld
to point; point	*zeigen; Punkt*
point of view	Standpunkt
to point out	hinweisen auf
to poison; poison	vergiften; Gift
Poland	Polen
Pole	Pole, Polin
police	*Polizei*
policeman	*Polizist*
policewoman	*Polizistin*
Polish	polnisch
polite	höflich
political	*politisch*
politician	*Politiker(in)*
politics	*Politik*
poll	Umfrage, Abstimmung, Wahl
to pollute	*verschmutzen*
polluted	*verschmutzt*
pollution	*Umweltverschmutzung*
poor	*arm*
popular	*beliebt, populär*
population	*Bevölkerung*
port	*Hafen*
portable	tragbar
positive	positiv
possession	Besitz
possibility	*Möglichkeit*
possible	*möglich*
post office	*Post*
postcard	*Postkarte*
postman	*Briefträger*
posture control	etwa: Körperhaltungsreglung

pot	*Topf*
potato	*Kartoffel*
potential	Potential, Möglichkeiten
poultry	Geflügel
pound	*Pfund*
pound note	Pfundnote
poverty	Armut
power	*Kraft*
power line	(Stark-) Stromleitung
power plant	Kraftwerk
power station	*Kraftwerk*
powered	angetrieben, betrieben
powerful	stark, mächtig
practical	praktisch
practice	*Übung*
to practise	*üben*
Prague	Prag
precise	genau
to predict	vorhersagen
predictable	vorhersagbar, vorhersehbar
prediction	Vorhersage
to prefer	*vorziehen*
prefix	Vorsilbe
preparation	Vorbereitung
to prepare	*vorbereiten*
to present; present	*schenken, vorstellen; Geschenk, Gegenwart; anwesend, gegenwärtig*
to press; press	*drücken, pressen; Presse*
pressure	Druck
pretty	*hübsch*
previous	vorhergehend, früher
price	*Preis*
pride	*Stolz*
primary school	Grundschule
prime energy	Primärenergie
prime minister	Premierminister(in)
principle	Grundsatz, Prinzip
to print	drucken
printer	Drucker(in)
printing	Druckwesen
private	privat
prize	*Preis, Gewinn*
probable	*wahrscheinlich*
process	Vorgang, Prozess
to produce	*produzieren, herstellen*
producer	*Hersteller(in), Produzent(in)*
product	*Produkt*
production	*Herstellung, Produktion*
production line	Fertigungsstraße, Fließband
productivity	Produktivität, Leistungsfähigkeit
professional	beruflich
to profit; profit	*profitieren; Gewinn*
profitable	*rentabel, einträglich, lohnend*
program(me)	*Programm*
progress	Fortschritt

project	Projekt
projection	Prognose, Projektion
proliferation	Ausbreitung, Umsichgreifen
to promise; promise	*versprechen; Versprechen*
to promote	(be)fördern, Werbung machen für
promotion	Werbung, Beförderung
to pronounce	*aussprechen*
pronounced	ausgeprägt, deutlich
pronunciation	*Aussprache*
proportion	Verhältnis, Teil, Anteil
proportional representation	Verhältniswahl
to prosper	gedeihen
prosperous	erfolgreich, florierend
to protect	*schützen*
protection	*Schutz*
protein	Protein
prototype	Prototyp
proud	*stolz*
to provide	sorgen für, liefern, bereitstellen
psychology	Psychologie
pub	*Kneipe*
public	*Öffentlichkeit; öffentlich*
public library	Stadtbücherei
public service	öffentlicher Dienst
to publish	veröffentlichen
publishing	Verlagswesen
to pull	*ziehen*
to pump; pump	pumpen; Pumpe
to punch; punch	schlagen; Schlag
punctual	pünktlich
to punish	(be)strafen
punishing	mörderisch, hart
pupil	*Schüler(in)*
purpose	*Zweck, Absicht*
to push; push	*schieben, stoßen; Stoß*
to put, (put, put)	*setzen, stellen, legen*
to put into practice	in die Praxis umsetzen, realisieren
to put on	*aufsetzen*
to put (s.o./sth.) to the test	jmdn./etwas auf die Probe stellen
to put up	*(Hand) heben, bauen, aufstellen*
puzzle	Rätsel
pylon	Mast

Q

qualification	Qualifikation, Befähigung
quality	*Qualität*
quantity	*Menge*
quantum	Quantum
quarter	*Viertel*
queen	*Königin*
question	*Frage*
quick	*schnell*
quiet	*ruhig*
to quit	verlassen, aufhören, aufgeben
quite	*ganz, ziemlich*
quotation	Zitat, Kostenvoranschlag

R

race	Rasse
racing bicycle	Rennrad
rack	Ständer, Regal
radical	radikal
rail	Schiene
railway	*Eisenbahn*
to rain; rain	*regnen; Regen*
to raise	*(hoch)heben, erheben,*
range	Sortiment, Umfang, Reichweite, Reihe
rape seed	Rapssamen
rapid	schnell
rate	Rate, (Prozent-)Satz
rather	*ziemlich*
raw	Roh-, roh
raw sewage	ungeklärte Abwässer
re-route	Umleitung
to re-use	wiederverwenden
to reach; reach	*erreichen; Reichweite*
to react	reagieren
reaction	Reaktion
to read, (read, read)	*lesen*
reader	*Leser(in)*
ready	*fertig, bereit*
real	*wirklich*
reality	*Wirklichkeit*
to realize	erkennen, bemerken
reason	*Grund*
reasonable	*vernünftig, günstig*
to recalculate	neu berechnen
to receive	*erhalten, bekommen*
recent	*jüngst, kürzlich*
recently	kürzlich
reception	*Empfang*
receptionist	*Empfangschef, Empfangsdame*
recession	Flaute
to recharge	(neu) aufladen
receipe	Rezept
recognition	Erkennung, Wiedererkennung
to recognize	erkennen, wiedererkennen
recommendation	Empfehlung
to record; record	*aufnehmen; Rekord, Aufnahme*
recording	(Ton-)Aufnahme
recording instruction	Aufnahmeanleitung
to recover	sich erholen
recovery	Erholung
rectangular	rechteckig
recyclable	wiederverwertbar
to recycle	wiederverwerten, wiederverwenden
red	*rot*

to reduce	verringern, senken
reduction	Verminderung
to refer to	(sich) beziehen auf
reference	Referenz, Bezug
with reference to	mit Bezug auf
to reflect	widerspiegeln
to refrigerate	im Kühlschrank aufbewahren, kühlen
to refuse	ablehnen, verweigern
region	Gebiet, Bezirk
to register	registrieren, anmelden,
regular	*regelmäßig*
to regulate	regulieren
regulation	Vorschriften, Regelungen
related	verwandt
relation	Beziehung
relationship	Verhältnis, Beziehung
relative	*Verwandte(r); relativ*
to relax	*(sich) entspannen*
to release	freilassen, veröffentlichen
to relocate	(Geschäftssitz) verlegen, umziehen
to rely on	abhängig sein von, sich verlassen auf
to remain	*bleiben*
to remark; remark	bemerken; Bemerkung
to remember	*sich erinnern an*
to remind (of)	*erinnern (an)*
reminder	Erinnerung, Mahnung
remote	entfernt, abgelegen
remote control	Fernbedienung
to remove	entfernen
renewable	erneuerbar
to rent; rent	*mieten, vermieten; Miete*
to repair; repair	*reparieren; Reparatur*
to repeat; repeat	*wiederholen; Wiederholung*
repetitive	(sich) wiederholend, eintönig
to replace	*ersetzen*
to reply; reply	*antworten; Antwort*
to report; report	*berichten; Bericht*
reported speech	indirekte Rede
to represent	darstellen, stehen für
to require	benötigen, erfordern
requirement	Bedarf, Erfordernis
to reserve	reservieren
to rescue; rescue	*retten; Rettung*
research	Forschung
researcher	Forscher(in)
research department	Forschungsabteilung
to reset	zurückstellen
to resign	zurücktreten
resolution	Auflösung
resort	Urlaubsort
resources	Boden-, Naturschätze, Mittel
responsibility	Verantwortung
responsible	verantwortlich
to represent	darstellen, stehen für
to rest; rest	*ruhen; Rest, Ruhe*
to result; result	*resultieren; Ergebnis*
retailer	Einzelhändler(in)
to return; return	*zurückkommen; Rückkehr, Rückgabe*
returnable	Mehrweg-
revival	Wiederaufleben, (-aufblühen)
to rewind	zurückspulen
to rewrite	neu schreiben, umschreiben
rich	*reich*
to ride, (rode, ridden); ride	*reiten, fahren; Ritt, Fahrt*
rice	Reis
right	*rechts*
to ring, (rang, rung); ring	*klingeln, anrufen; Ring, Anruf*
riot	Aufruhr, Krawall
to rise, (rose, risen); rise	*(auf)steigen, zunehmen; Aufstieg, Zunahme*
to risk; risk	riskieren; Risiko
river	*Fluss*
road	*Straße*
road works	Straßenbauarbeiten
to roar away	„wegdonnern"
robot	Roboter, vollautomatische Vorrichtung
role	Rolle
to roll	rollen
romantic	romantisch
roof	*Dach*
room	*Zimmer*
rope	Seil
to rot	verrotten
rotation	Drehung
round	*Runde; rund, herum*
routinely	routinemäßig
rubber	*(Radier-)Gummi*
rubbish	*Abfall, Müll*
to ruin	ruinieren
to rule; rule	*herrschen; Regel, Herrschaft*
to run, (ran, run)	*laufen*
to run on	laufen mit
run-down	heruntergekommen
rural	ländlich
to rush (s.o. to)	eilen, jmdn./etwas schnell an einen Ort bringen
rush-hour	Hauptverkehrszeit
Russia	Russland

S

to sacrifice; sacrifice	opfern; Opfer
sad	*traurig*
safe	*sicher*
safety	*Sicherheit*
salary	Gehalt
sale	*(Schluss-)Verkauf*
sales	Absätze
sales figures	Verkaufszahlen
sales department	Verkaufsabteilung
salesman/woman	*Verkäufer(in)*
salt	*Salz*
same	*gleiche(r,s), der-, die-, dasselbe*
satisfaction	Zufriedenheit
satisfied	zufrieden
sausage	Wurst
to save	*retten*
to say, (said, said)	*sagen*
scale	Maß(-stab)
scanner	Scanner, Abtaster
scarcity	Knappheit
sceptical	skeptisch
schedule	(Zeit-, Fahr-) Plan, Programm
scheme	Projekt, Plan, Programm
school	*Schule*
scientific	wissenschaftlich
scientist	Wissenschaftler(in)
to score; score	(Treffer) erzielen; Spielstand
scramble	Gerangel, Gedrängel
scrap	Schrott, Fetzen
scrap heap	Müll-, Schutt-, Schrotthaufen
screen	Bildschirm, Leinwand
to screw; screw	schrauben; Schraube
sea	*Meer*
seal	Seehund
seaside resort	Seebad
seat	*Sitz*
second	*Sekunde; zweite(r,s)*
secret	*Geheimnis; geheim*
secretary	*Sekretär(in)*
secure	sicher
to see, (saw, seen)	*sehen*
to seem	*scheinen*
seemingly	offensichtlich, scheinbar
to select	auswählen
self-employed	selbstständig
to sell, (sold, sold)	*verkaufen*
seller	*Verkäufer(in)*
seminar	Seminar
to send, (sent, sent)	*schicken, senden*
sensitive	empfindlich, heikel
sensor	Sensor, Fühler
to sentence; sentence	*verurteilen; Satz, Urteil*
to separate; separate	trennen; getrennt
series	Reihe, Folge
serious	*ernst*
to serve	*(be)dienen, servieren*
serve	Aufschlag (Tennis)
to service; service	*(Auto) warten; Dienst, Betrieb*
service industry	*Dienstleistungsbranche*

to set, (set, set)	*setzen, stellen, legen*
to sever	durchtrennen, abbrechen
several	*mehrere*
severe	ernst
sewage	Abwasser
to shake (off), (shook, shaken)	(ab)schütteln
shall	*sollen*
to share; share	*teilen; Anteil*
sharp	scharf, deutlich, steil
sheet	*Blatt, Bogen*
shelf	*Regal*
to shift	verlagern
to shine, (shone, shone)	*scheinen, leuchten*
to ship; ship	*verschicken; Schiff*
shipbuilding	Schiffbau
shipyard	Werft
shirt	*Hemd*
shoe	*Schuh*
shop	*Geschäft, Laden*
shop floor	Produktionsstätte, -halle
shop assistant	*Verkäufer(in)*
shopper	Käufer(in)
shopping	Einkauf(en)
short	*kurz*
shortage	Mangel, Knappheit
short cut	Abkürzung
short-range radio	Kurzwelle
short-term	kurzfristig
to shorten	kürzen
should	*sollte(n)*
shoulder	*Schulter*
to shout; shout	*rufen; Ruf*
to show, (showed, shown); show	*zeigen; Ausstellung, Vorstellung*
shower	Dusche
to shut, (shut, shut)	*schließen*
sick	*krank*
sick leave	krankheitsbedingte Abwesenheit
side	*Seite*
side impact beam/ protection	Seitenaufprallschutz
sight	*Sicht, Anblick, Sehenswürdigkeit*
to sign; sign	*unterschreiben; Zeichen*
significant	bedeutend, bedeutsam
silencer	Schalldämpfer
silicon chip	Siliziumchip
similar	ähnlich
simple	*einfach*
simple majority voting system	Mehrheitswahl-system
to simulate	simulieren
since	*seit*
since then	*seitdem*
to sing, (sang, sung)	*singen*

singer	Sänger(in)
single	einzig, hier: einheitlich
to sink; sink	sinken; Waschbecken
Sir	*Herr (Anrede)*
sister	*Schwester*
to sit down	*sich hinsetzen*
to sit, (sat, sat)	*sitzen*
site	Grundstück, Sitz, Standort
size	*Größe*
skill	*Fertigkeit, Fähigkeit, Können*
skilled	qualifiziert, ausgebildet
skirt	*Rock*
sky	*Himmel*
skyscraper	Wolkenkratzer
to sleep, (slept, slept); sleep	*schlafen; Schlaf*
slight	leicht, schwach, gering
slim	*schlank*
slogan	Slogan, Werbespruch
slope	Abhang, Gefälle, Schräge
slow	*langsam*
to slow down	*verlangsamen*
slow moving traffic	zäh fließender Verkehr
small	*klein*
smart house	intelligentes Heim
to smell; smell	*riechen; Geruch*
to smile; smile	*lächeln; Lächeln*
to smoke; smoke	*rauchen; Rauch*
smoker	Raucher(in)
smokestack	Schornstein
to snap	einrasten
to snow; snow	*schneien; Schnee*
so far	*bisher, bis jetzt*
so-called	so genannt
soap	*Seife*
to soar	(stark) ansteigen
soccer	*Fußball*
social	gesellschaftlich, sozial
social studies	Sozialwissen-schaften
society	Gesellschaft
sociology	Soziologie
sock	*Socke*
soft	*weich*
soil	Boden, Erde
solar power	Sonnenenergie
soldier	Soldat(in)
solid	massiv, verlässlich, ganz
solitary	einsam
solution	*Lösung*
to solve	*lösen*
some	*einige*
somebody	*jemand*
someone	*jemand*
something	*etwas*
sometimes	*manchmal*
somewhere	irgendwo, ungefähr
son	*Sohn*

song	*Lied*
soon	*bald*
sophisticated	hochentwickelt, anspruchsvoll, ausgeklügelt
sorry	*betrübt, tut mir Leid!, Entschuldigung!*
sort	Art, Sorte
to sound; sound	*klingen; Geräusch, Klang*
sound wave	Schallwelle
source	Quelle
south	*Süden, südlich, Süd-, südwärts*
southern	Süd-, südlich
sovereign	Souverän, Herrscher
space	*Raum, Platz, Weltraum*
Spain	Spanien
to span; span	spannen; Spanne
spare time	*Freizeit*
spark	(Zünd-)Funke
spark plug	Zündkerze
to speak, (spoke, spoken)	*sprechen*
speaker	*Sprecher(in)*
special	*besondere(r,s), spezielle(r,s)*
specialist	Spezialist(in)
species	(Tier-,Pflanzen-)Art, Spezies
spectacular	spektakulär
speech	*Rede*
to speed; speed	*schnell fahren; Geschwindigkeit*
to speed up	*beschleunigen*
speedy	schnell
to spell; spell	*buchstabieren; Zauber(spruch)*
to spend, (spent, spent)	*ausgeben, verbringen*
spending power	Kaufkraft
to spin	(sich) drehen
spirit	Geist(eshaltung), Sinn, Einstellung
to spoil	verderben, ruinieren
spokesman	Sprecher
spoon	*Löffel*
sports	Sport
sportsman	Sportler
sportswoman	Sportlerin
to spray; spray	sprühen, spritzen, lackieren; Spray
to spread	sich verbreiten
spring	*Frühling*
square	Quadrat
to squeeze	sich drängen, quetschen
stable	stabil, dauerhaft
staff	*Belegschaft, Kollegium*
staffroom	Lehrerzimmer
stage	Phase, Stadium, Bühne
stamp	*Briefmarke*

to stand, (stood, stood); stand	*stehen, aushalten; Stand*
standard	Niveau, Standard, Maßstab
standard of living	Lebensstandard
to standardize	standardisieren
standstill	Stillstand
to stare	starren
to start; start	*anfangen; Anfang*
to start up (business)	(Geschäft) anfangen
to state; state	*erklären, darlegen; Staat, Zustand*
station	*Bahnhof*
statistics	Statistik
to stay	*bleiben*
steady	stabil, gleichbleibend
to steal	stehlen
steam	Dampf
steel	Stahl
steel support	Stahlträger
steering wheel	Lenkrad
step	Schritt, Stufe
still	*dennoch, noch*
stock market	Börse
stone	*Stein*
to stop; stop	*(an)halten; Halt*
storage	Lagerung, Aufbewahrung, Speicherung
to store; store	*lagern, speichern; Laden (AE), Kaufhaus, Lager*
storm	*Sturm*
story	*Geschichte*
straight in	(gerade) direkt hinein
to strain; strain	(sich) anstrengen; Belastung, Anstrengung
strange	*merkwürdig*
strap	Riemen, Träger, Bandage
strategy	Strategie
stream	Bach, Wasserlauf
street	*Straße*
strength	Stärke
to strike; strike	*streiken, schlagen; Streik*
striker	Streikende(r)
stroke	Schlag
strong	*stark*
to struggle with	kämpfen mit
structure	Struktur
studies	Kurs, Studium
to study; study	*studieren; Studium*
style	Stil, Art
sub-station	Umspannungswerk
subject	*Fach, Thema*
subject matter	Stoff, Inhalt
substantial	beträchtlich, erheblich
to substitute; substitute	ersetzen; Ersatz
substitution	Ersetzen, Austausch
suburb	*Vorort*
to succeed (in)	*Erfolg haben (in, bei)*

success	*Erfolg*
successful	*erfolgreich*
such	*solche(r,s)*
such as	wie (zum Beispiel)
to suck	(an)saugen
sudden	*plötzliche(r,s)*
to suffer from	leiden unter
sufficient	ausreichend
sugar	*Zucker*
to suggest	vorschlagen
suggestion	Vorschlag
to suit; suit	*passen; Anzug*
suitable	passend, geeignet
suitcase	*Koffer*
sum	*Summe*
to sum up	zusammenfassen
summary	Zusammenfassung
summer	*Sommer*
sun	*Sonne*
sunchair	Liegestuhl
sunshine	*Sonnenschein*
superb	erstklassig
superfast	außerordentlich schnell
supermarket	*Supermarkt*
supper	*Abendessen*
supplier	Lieferant(in)
to supply; supply	versorgen, liefern; Versorgung
to suppose	*annehmen, vermuten*
sure	*sicher*
surface	Oberfläche
surgery	Chirurgie
to surprise; surprise	*überraschen; Überraschung*
surprised	*überrascht*
survey	*Überblick, Umfrage*
survival	Überleben
to survive	überleben
suspender	Aufhängevorrichtung, Hosenträger
suspension	Aufhängung, Ausschluss
suspension bridge	Hängebrücke
suspicious of	misstrauisch gegenüber
to swallow	schlucken
sweet	*süß*
to swim, (swam, swum)	*schwimmen*
to switch; switch	*schalten, wechseln; Schalter*
syndrome	Syndrom

T

table	*Tisch, Tabelle*
to take, (took, taken)	*nehmen*
to take a test	einen Test machen
to take acton	etwas unternehmen
to take away	wegnehmen
to take care	aufpassen
to take notes	Notizen machen
to take off	abnehmen, starten (Flugzeug)
to take over	übernehmen
to take part in	*teilnehmen an*
to take place	*stattfinden*
to take the view	der Ansicht sein
to take to pieces	zerlegen
to talk; talk	*sprechen, reden; Gespräch*
tall	*hoch(gewachsen), groß*
tape	Band, Klebstreifen
target group	Zielgruppe
task	Aufgabe
to taste; taste	*probieren, schmecken; Geschmack*
tasty	*lecker*
tax	*Steuer*
tea	*Tee*
to teach, (taught, taught)	*lehren, unterrichten*
teacher	*Lehrer(in)*
team	*Mannschaft*
team work	Teamarbeit
technical	*technisch*
technique	Technik
technological	technologisch, technisch
technologist	Technologe, Technologin, Techniker(in)
technology	*Technik, Technologie*
telephone	*Telefon*
television (TV)	*Fernsehen*
to tell, (told, told)	*erzählen*
temperature	Temperatur
to tend	tendieren, neigen
tense	Zeitform
tension	Spannung
tent	Zelt
term	Ausdruck, Wort
terrible	*schrecklich*
to test; test	*testen, überprüfen untersuchen; Klassenarbeit, Versuch*
textile	Stoff
than	*als*
to thank; thank	*danken; Dank*
thanks to	wegen, dank
that	*dass, jene(r,s), welche(r,s)*
then	*dann, damals*
theory	Theorie
there	*dort*
there is/are	*es gibt*
therefore	*deshalb*
these	*diese*
these days	*heutzutage*
thick	*dick*
to thicken	dicker werden, dicker machen
thin	*dünn*
thing	*Ding*

to think, (thought, thought)	*denken*
thirsty	*durstig*
this	*diese(r,s)*
thorough	gründlich
those	*jene*
though	*obwohl, trotzdem*
thought	Gedanke
thousand	*Tausend*
thousandth	Tausendstel
threat	(Be-)Drohung
to threaten	(be)drohen
through	*durch*
throughout	die ganze Zeit, stets
to throw, (threw, thrown)	*werfen*
thus	*so*
ticket	*Fahrkarte*
tides	Gezeiten
tidy	*ordentlich, aufgeräumt*
to tie	binden, bündeln
to tighten	(fest) anziehen, verschärfen
till	*bis*
to tip	*kippen*
time	*Zeit*
times	*Mal(e)*
tired	*müde*
title	Titel
to transport; transport	*transportieren; Transport*
tobacco	*Tabak*
today	*heute*
toe	*Zeh*
together	*zusammen*
toilet	*Toilette*
tomato	*Tomate*
tomorrow	*morgen*
ton(ne)	Tonne
tongue	*Zunge*
tonight	*heute Abend*
too	*auch*
tool	*Werkzeug*
tooth, teeth	*Zahn, Zähne*
top	*Spitze*
topic	Thema
total	gesamt
to touch; touch	berühren; Berührung
tough	streng, zäh
tour	*(Rund-)Reise, Tour*
tourism	Tourismus
tourist board	Fremdenverkehrsamt
tourist guide	Fremdenführer(in)
toward(s)	*in Richtung*
tower	*Turm*
town	*Stadt*
town hall	Rathaus
toxic	giftig
toy	*Spielzeug*
to trace; trace	nachgehen, nachspüren; Spur
trading	Handel
to trade (in); trade	*handeln (mit); Gewerbe, Handwerk, Handel*
trading centre	Handelszentrum

traditional	*traditionell, herkömmlich*
traffic	*Verkehr*
traffic flow	Verkehrsfluss
traffic jam	Verkehrsstau
traffic light(s)	*Ampel*
to train; train	*ausbilden, eine Ausbildung machen, trainieren; Zug*
training	*Ausbildung, Training*
tram	Straßenbahn
tramway	Straßenbahngleis
to transfer	übertragen
transformer	Umwandler
transgenic	Gene von anderen Lebewesen beinhaltend
to translate	*übersetzen*
translation	*Übersetzung*
transmission	Übertragung
transmission cable	Leitungskabel
to transmit	übermitteln
to travel	*reisen, fahren*
to treat	behandeln
treatment	Behandlung
treaty	Vertrag
tree	*Baum*
trial and error	Ausprobieren, Versuch und Irrtum
trillion	Trillion (AE: Billion)
trip	*(Kurz-)Reise*
tropical	tropisch
trouble	*Schwierigkeit(en)*
trousers	*Hose*
true	*wahr, richtig*
truth	*Wahrheit*
to try; try	*versuchen; Versuch*
turbine	Turbine
to turn	*(sich) drehen, wenden*
to turn into	umwandeln in
to turn left/right	*links/rechts abbiegen*
to turn on/off	*ein-/ausschalten*
to turn over	*umdrehen, überschlagen*
to turn to	sich zuwenden
turnover	Umsatz
TV set	Fernsehgerät
twice	*zweimal*
twin-engined	zweimotorig
twin towers	Zwillingstürme
two-legged	zweibeinig
to type; type	*(Maschine)schreiben, tippen; Art, Typ*
typewriter	*Schreibmaschine*
typical	*typisch*
typist	*Schreibkraft*

U

ugly	hässlich
ultimate	vollendet, perfekt
umbrella	*Schirm*
unable	nicht in der Lage (sein)
unavoidable	unvermeidbar
unbelievable	unglaublich
uncle	*Onkel*
undemocratic	undemokratisch
underground	Untergrund, U-Bahn
to underline	unterstreichen
underpaid	unterbezahlt
to understand, (understood, understood)	*verstehen*
undesirable	unerwünscht
unemployed	*arbeitslos*
unemployment	*Arbeitslosigkeit*
unemployment rate	Arbeitslosenquote
unemployment benefit	Arbeitslosenunterstützung
unfortunate	*unglücklich*
unfortunately	*leider*
unfriendly	*unfreundlich*
unhappy	*unglücklich*
unimportant	unwichtig
uninhabitable	unbewohnbar
union	Gewerkschaft
unit	Einheit, Gruppe, Kapitel
united	vereinigt
United Kingdom	Vereinigtes Königreich
unity	Einheit
university	*Universität*
unknown	unbekannt
unless	wenn … nicht, es sei denn
unlikely	unwahrscheinlich
unlimited (company)	Gesellschaft mit unbeschränkter Haftung
to unlock	aufschließen
unpopular	unpopulär
unreliable	unzuverlässig
unsuitable	unpassend, ungeeignet
unsupported	ungestützt, nicht unterstützt
until (till)	*bis*
unusual	ungewöhnlich
unwilling	nicht bereit (etwas zu tun), widerwillig
up to	bis zu
up to now	bis jetzt
up until now	bis jetzt
upon	*auf*
upper	obere(r,s)
upstairs	*(nach) oben, im Obergeschoss*
urban	städtisch, Stadt-
urbanisation	Verstädterung
to urge; urge	drängen; Drang
to use; use	*gebrauchen; Gebrauch*
used	*gebraucht, gewohnt*
useful	*nützlich*
useless	*nutzlos*
user	Benutzer(in)
user-friendly	benutzerfreundlich
usual	*gewöhnlich*

V

vacancy	freie Stelle
vacation	*Ferien (AE)*
vaccine	Impfstoff
vacuum-cleaner	Staubsauger
valley	Tal
value	Wert
van	*Lieferwagen*
various	verschieden, vielfältig
to vary	sich ändern, unterschiedlich sein
vegetable(s)	*Gemüse*
vehicle	*Fahrzeug*
version	Version, Fassung
very	*sehr*
via	über
video production	Videoproduktion
Vienna	Wien
Vietnamese	Vietnamese, Vietnamesin; vietnamesisch
view	Sicht, Ansicht, Aussicht
viewer	Zuschauer(in)
viewpoint	Standpunkt
village	*Dorf*
virtual	virtuell
visa	Visum
to visit; visit	*besuchen; Besuch*
visitor	*Besucher(in)*
visual	visuell, optisch
vital	lebenswichtig, entscheidend
vocabulary	*Wortschatz, Vokabelverzeichnis*
voice	*Stimme*
voice recognition	Spracherkennung
voltage	elektrische Spannung
volume	Lautstärke
to vote; vote	*abstimmen, wählen; Abstimmung, Stimme*
voter	Wähler(in)
voting system	Wahlsystem

W

wage	Lohn
to wait	*warten*
waiter	*Kellner*
waitress	*Kellnerin*
to wake (up), (woke, woken)	*wecken, aufwachen*
to walk; walk	*gehen; Spaziergang*
wall	*Wand, Mauer*
to want	*wollen*
war	*Krieg*
warehouse	*Lager*
to warm; warm	*wärmen; warm*
to warn	*warnen*
warning	Warnung
Warsaw	Warschau
wartime	Kriegszeit
to wash	*waschen*
washing machine	*Waschmaschine*
to waste; waste	*verschwenden; Abfall, Verschwendung*
waste management	Abfallwirtschaft
waste-melting	„Einschmelzen" von Abfall
wasteful	verschwenderisch
to watch; watch	*beobachten, sehen; Uhr*
water	*Wasser*
water table	Grundwasserspiegel
wave	Welle
way	*Weg, Art und Weise*
way of life	Lebensweise
weak	*schwach*
to weaken	schwächen
wealth	Wohlstand, Reichtum
wealthy	wohlhabend
to wear, (wore, worn)	*(Kleidung) tragen*
weather	*Wetter*
wedding	Hochzeit
week	*Woche*
weekend	*Wochenende*
weight	*Gewicht*
to welcome; welcome	*willkommen (heißen); Willkommen, Empfang*
to weld	schweißen
welfare	Wohl, Wohlfahrt, Sozialhilfe
well	*gut, gesund; na ja ...*
well-known	*bekannt*
well-meaning	wohlmeinend
well-paid	gut bezahlt
Welsh	walisisch
west	*Westen, westlich, West-, westwärts*
western	West-, westlich
wet	*nass, feucht*
what	*was*
what about?	*was ist mit?*
what else	*was noch*
what for	*wofür, wozu*
what ... like?	wie ...?
wheel	*Rad*
when	*wann*
where	*wo*
whereas	während
whether	*ob*
which	*welche(r,s)*
while	*während*
white	*weiß*
who	*wer*
whole	*ganz*
whom	*wem, wen*
whose	*wessen*
why	*warum*
wide	*weit*
wife	*Ehefrau*
will	*Wille; werden*
willing	bereit
to win, (won, won)	*gewinnen*
window	*Fenster*
windscreen	Windschutzscheibe
windy	windig
wine	*Wein*
winner	Gewinner(in)
to wire up	anschließen, verkabeln
wired	verkabelt
wireless	drahtlos
wisdom	Weisheit, Kenntnisse
to wish; wish	*wünschen; Wunsch*
within	*innerhalb*
without	*ohne*
woman	*Frau*
to wonder; wonder	sich fragen, sich wundern; Wunder
wonderful	*wunderbar*
wood	*Holz, Wald*
wooden	hölzern
wool	*Wolle*
word	*Wort*
word-by-word	Wort für Wort
to work; work	*arbeiten; Arbeit*
to work out	herausfinden, ausarbeiten, lösen
worker	*Arbeiter(in)*
working class	Arbeiterklasse
working hours	Arbeitszeit
working week	Arbeitswoche
workmate	Kollege
workshop	Werkstatt
world	*Welt*
worldwide	weltweit
worried	besorgt
to worry	sich sorgen
worrying	besorgniserregend
worse	*schlechter*
worst	*am schlechtesten*
worth	*wert*
would like	würde(n) gerne
to write, (wrote, written)	*schreiben*
writer	*Verfasser(in), Schriftsteller(in)*
wrong	*falsch*

Y

year	*Jahr*
yearly	jährlich
(... years) ago	*vor (... Jahren)*
yellow	*gelb*
yesterday	*gestern*
yet	*jedoch, schon*
you're welcome	*gern geschehen!*
young	*jung*
Yours	Dein(e)
Yours faithfully	Mit freundlichen Grüßen
Yours sincerely	Mit freundlichen Grüßen
youth	*Jugend*
youth hostel	Jugendherberge

Z

zero	null
zinc	Zink

AE = American English

A	
a couple of	ein paar
a little	etwas, ein bisschen
a lot of	viel
a number of	einige
to be able	können
abroad	im/ins Ausland
to accept	annehmen, akzeptieren
accident	Unfall
accurate	genau
activity	Aktivität
to add	hinzufügen, addieren
address	Adresse
advantage	Vorteil
to advertise	werben, inserieren
advertisement	Werbeanzeige, Anzeige
advertising	Werbung
advice	Rat, Ratschlag
to advise	(be)raten
to be afraid (of)	Angst haben (vor)
afternoon	Nachmittag
afterwards	danach
again	wieder
against	gegen
age	Alter, Zeitalter
ago	vor
to agree	einverstanden sein, zustimmen
agreement	Übereinstimmung, Übereinkunft
air	Luft
airport	Flughafen
all	alle(s), ganz
to allow	erlauben
almost	fast, beinahe
alone	allein
along	entlang
already	schon
also	auch
alternative	Alternative, Wahl; alternativ
although	obwohl
always	immer
among	zwischen, unter
amount	Betrag, Summe
angry	wütend
animal	Tier
to announce	bekannt geben, ansagen
announcement	Bekanntgabe, Durchsage
another	noch eine(r,s), ein(e) andere(r,s)
to answer; answer	(be)antworten; Antwort
any	irgendeine(r,s)
any time	jederzeit
anybody	irgendjemand
anyone	irgendjemand
anything	irgendetwas
to appear	(er)scheinen
apple	Apfel
to apply (for)	sich bewerben (um)
area	Gebiet
around	herum, um, ungefähr
arrival	Ankunft
to arrive	ankommen
article	Artikel
as	wie, da, als
as well	auch, ebensogut
to ask (for)	fragen, bitten (um)
at first	anfangs
at home	zu Hause
at last	schließlich, endlich
at least	wenigstens
at once	sofort
at present	im Augenblick, jetzt
at the moment	im Moment, jetzt
at work	bei der Arbeit
to attend (school)	(Schule) besuchen
attention	Aufmerksamkeit
aunt	Tante
autumn	Herbst
away	weg
awful	furchtbar, schrecklich

B	
back	Rücken; zurück
bad	schlecht, schlimm, böse
bag	Tüte, Tasche
baggage	Gepäck
to bake	backen
baker	Bäcker(in)
balcony	Balkon
bank	Bank (Geldinstitut)
basket	Korb
bathroom	Badezimmer
beach	Strand
beard	Bart
beautiful	schön
because	weil
because of	wegen
to become, (became, become)	werden
bed	Bett
bedroom	Schlafzimmer
before	vorher, ehe
to begin, (began, begun)	beginnen
beginning	Anfang
behind	hinter
to believe	glauben
to belong to	gehören
best	am besten
best wishes	alles Gute
better	besser
between	zwischen
bicycle	Fahrrad
big	groß
bike	Fahrrad (Kurzform)
bill	Rechnung
bird	Vogel

birth	Geburt
birthday	Geburtstag
bit	Stückchen
to bite, (bit, bitten)	beißen
black	schwarz
blackboard	(Wand-)Tafel
to blame (for)	verantwortlich machen (für)
block of flats	Wohnblock
blood	Blut
blue	blau
board	Brett
boat	Boot
body	Körper
to boil	kochen
to book; book	buchen; Buch
boot	Stiefel
border	Grenze
bored	gelangweilt
boring	langweilig
to borrow	sich ausleihen
both	beide
bottle	Flasche
bottom	Boden
box	Schachtel, Karton, Kiste
boy	Junge
boyfriend	Freund
bread	Brot
to break, (broke, broken); break	(zer)brechen; Pause
to break down	zusammenbrechen
breakfast	Frühstück
bridge	Brücke
to bring, (brought, brought)	bringen
to broadcast, (broadcast, broadcast)	übertragen, senden
brochure	Broschüre, Prospekt
brother	Bruder
brown	braun
to brush; brush	bürsten; Bürste
to build, (built, built)	bauen
building	Gebäude
to burn, (burnt, burnt)	brennen
business	Geschäft
busy	beschäftigt
but	aber
butcher	Metzger(in)
to buy, (bought, bought)	kaufen
buyer	Käufer(in)
by	durch, bei, neben, bis
by the way	übrigens
C	
cake	Kuchen
to calculate	(be)rechnen
to call; call	(an)rufen, nennen; Anruf
camera	Kamera
to camp; camp	zelten, lagern; Lager
can	können; Büchse, Dose
capital	Großbuchstabe, Hauptstadt
car	Auto

card	Karte
to care; care	sorgen; Sorge
career	Beruf, Laufbahn, Karriere
careful	sorgfältig
careless	sorglos
cargo	Fracht, Ladung
to carry	tragen
case	Fall, Koffer, Kiste
cash	Bargeld
cat	Katze
to catch, (caught, caught)	fangen
to cause; cause	verursachen; Ursache
ceiling	Zimmerdecke
centre	Zentrum, Mittelpunkt
century	Jahrhundert
certain	sicher
chair	Stuhl
chalk	Kreide
to change; change	wechseln, (sich) (ver)ändern; Wechsel(geld), Änderung
channel	Kanal
chart	Schaubild, Tabelle, Karte
cheap	billig
to check; check	kontrollieren; Kontrolle
cheers	Tschüs, zum Wohl
cheese	Käse
child	Kind
chips	Pommes Frites, Kartoffelchips
chocolate	Schokolade, Praline
choice	Wahl
to choose, (chose, chosen)	(aus)wählen
Christmas	Weihnachten
church	Kirche
cigarette	Zigarette
cinema	Kino
city	Stadt
class	Klasse
classmate	Klassenkamerad(in)
classroom	Klassenzimmer
to clean; clean	reinigen; sauber
to clear; clear	säubern, räumen; klar
clever	klug
to climb	klettern
clock	Uhr
to close; close	schließen; dicht, nahe
clothes	Kleidung
cloud	Wolke
coal	Kohle
coast	Küste
coat	Mantel
coffee	Kaffee
coin	Münze
cold	kalt
to collect	sammeln, abholen
collection	Sammlung
college	College, Fach(hoch)schule
colour	Farbe
to comb; comb	kämmen; Kamm
to come, (came, come)	kommen
comfortable	bequem
to communicate	kommunizieren, übermitteln
communication	Kommunikation, Übermittlung

company	Gesellschaft, Firma
to compare	vergleichen
competition	Wettbewerb, Konkurrenz
competitor	Konkurrent(in), Wettbewerbsteilnehmer(in)
to complete; complete	vervollständigen; vollständig
to connect	verbinden
connection	Verbindung, Zusammenhang
to contact; contact	sich in Verbindung setzen mit; Verbindung, Kontakt
to contain	enthalten
container	Behälter
to continue	fortsetzen, weitermachen, weitergehen
to cook; cook	kochen; Koch, Köchin
cooker	Herd
to cooperate	zusammenarbeiten
cooperation	Zusammenarbeit
to copy; copy	kopieren, abschreiben; Kopie, Exemplar
corner	Ecke
to correct; correct	korrigieren, verbessern; richtig
correction	Verbesserung, Korrektur
to cost, (cost, cost); cost	kosten; Kosten
cotton	Baumwolle
to count	zählen, ins Gewicht fallen
counter	Ladentisch, Schalter
country	Land
couple	Ehepaar, Paar
course	Kurs, Lauf, Gang, Strecke
to cover; cover	(be)decken; Abdeckung, Umschlag
cow	Kuh
to cross	überqueren
crossing	Kreuzung
crowd	Menschenmenge, Menge
crowded	dicht gedrängt, voll
to cry; cry	schreien, weinen; Schrei
cup	Tasse, Pokal
curious	neugierig, seltsam
curtain	Vorhang
customer	Kunde, Kundin
customs	Zoll, Zoll-
to cut, (cut, cut); cut	schneiden; Schnitt
to cycle	Fahrrad fahren
D	
daily	täglich
to damage; damage	beschädigen; Schaden
to dance; dance	tanzen; Tanz
danger	Gefahr
dangerous	gefährlich
dark	dunkel
darkness	Dunkelheit
date	Datum, Zeitpunkt, Verabredung
date of birth	Geburtsdatum
daughter	Tochter
day	Tag
dead	tot
to deal, (dealt, dealt) (in, with); deal	handeln (mit, von); Geschäft

dear	lieb
death	Tod
to decide	(sich) entscheiden
decision	Entscheidung
deep	tief
to deliver	liefern, (Rede) halten
delivery	Lieferung
department	Abteilung
department store	Kaufhaus
to depend (on)	abhängen (von)
to describe	beschreiben
description	Beschreibung
to design; design	entwerfen, konstruieren; Entwurf, Muster
desk	Schreibtisch
to destroy	zerstören
destruction	Zerstörung
detail	Einzelheit, Detail
to develop	entwickeln
development	Entwicklung
to dial	(Telefon) wählen
to dictate	diktieren
dictation	Diktat
dictionary	Wörterbuch
to die	sterben
to differ (from)	sich unterscheiden (von)
difference	Unterschied
different	unterschiedlich
difficult	schwierig
difficulty	Schwierigkeit
dining room	Esszimmer
dinner	(Mittag-, Abend-)Essen
to direct (to); direct	richten, lenken (auf); direkt
direction	Richtung
dirty	schmutzig
disadvantage	Nachteil
to disagree (with)	anderer Meinung sein, nicht übereinstimmen mit
to discuss	besprechen, diskutieren
discussion	Gespräch, Unterredung, Diskussion
to dislike	nicht mögen
distance	Entfernung, Strecke, Abstand
doctor	Arzt, Ärztin
dog	Hund
door	Tür
down	hinunter, unten
downstairs	die Treppe hinunter, unten, im Erdgeschoss
to draw	zeichnen, ziehen
drawing	Zeichnung
to dream; dream	träumen; Traum
to dress; dress	(sich) anziehen; Kleid, Kleidung
to drink, (drank, drunk); drink	trinken; Getränk
to drive, (drove, driven); drive	fahren, (an)treiben; Fahrt
driver	Fahrer(in)
to dry; dry	trocknen; trocken
during	während
dustbin	Mülltonne
duty	Pflicht, Zoll

E	
each	jede(r,s)
each other	einander, sich
ear	Ohr
early	früh
to earn	verdienen
earth	Erde
east	Osten, östlich, Ost-, ostwärts
easy	leicht
to eat, (ate, eaten)	essen
economy	Wirtschaft
edge	Kante, Rand, Schneide
to educate	erziehen
education	Erziehung
to effect; effect	durchführen, erzielen, leisten; Wirkung
egg	Ei
either ... or	entweder ... oder
to elect	wählen
election	Wahl
electric	elektrisch
electricity	Elektrizität
electronics	Elektronik
else	sonst
emergency	Notfall
to employ	beschäftigen, einstellen
employed	beschäftigt
employee	Angestellte(r), Arbeitnehmer(in)
employer	Arbeitgeber(in)
employment	Beschäftigung, Anstellung, Arbeit
empty	leer
to end; end	(be)enden; Ende
energy	Energie, Kraft
engine	Motor, Triebwerk, Lokomotive
engineer	Ingenieur(in), Techniker(in)
engineering	Technik, technisch
to enjoy	genießen, sich freuen an
enjoyable	schön, angenehm, unterhaltsam
enough	genug
to enquire (about)	sich erkundigen (nach), fragen (nach)
enquiry	Anfrage, Erkundigung, Untersuchung
to enter	eintreten
entrance	Eintritt, Eingang
envelope	Umschlag
environment	Umwelt
equal	gleich
equipment	Ausrüstung, Ausstattung
especial	besondere(r,s)
even	sogar, selbst; eben, gleich
evening	Abend
event	Ereignis
ever	je
every	jede(r,s)
everybody	jeder
everyone	jeder
everything	alles
everywhere	überall
exact	genau
exam(ination)	Prüfung
example	Beispiel
excellent	ausgezeichnet, hervorragend
exciting	aufregend
to excuse; excuse	(sich) entschuldigen; Entschuldigung
exercise	Übung
to exist	existieren, bestehen
exit	Ausgang, Ausfahrt
to expect	erwarten
expensive	teuer
experience	Erfahrung
to explain	erklären
explanation	Erklärung
to export; export	exportieren, ausführen; Ausfuhr, Export
to express; express	ausdrücken, äußern; Schnellzug, als Eilsache
expression	Ausdruck
extra	besonders
eye	Auge

F	
to face; face	(einer Sache) ins Auge sehen, gegenübertreten; Gesicht
fact	Tatsache
factory	Fabrik
fair	Messe, Markt; gerecht, fair
to fall, (fell, fallen); fall	fallen; Fallen, Sturz; (AE) Herbst
false	falsch
family	Familie
famous	berühmt
far	weit
fare	Fahrpreis, Fahrgeld
farm	Bauernhof, Farm
farmer	Landwirt
farming	Landwirtschaft
fast	schnell
fat	Fett; dick
father	Vater
favourite	Lieblings-; Liebling
to feed, (fed, fed)	füttern
to feel, (felt, felt)	fühlen
feeling	Gefühl
few	wenige
field	Feld
to fight, (fought, fought); fight	kämpfen; Kampf
to fill	füllen
final	Finale, Endrunde; letzte(r,s), endgültig
to find, (found, found)	finden
fine	Strafe; fein, gut, schön
to finish; finish	beenden; Ziel, Vollendung
to fire; fire	feuern; Feuer
firm	Firma; fest, verbindlich
first	erste(r,s), zuerst
first of all	zuerst, vor allem
to fish; fish	fischen; Fisch

to fit; fit	(zusammen) passen, einbauen; gesund, in Form
to fix	befestigen, reparieren, besorgen
flat	Wohnung; flach
flight	Flug
floor	Boden
flower	Blume
to fly, (flew, flown); fly	fliegen; Fliege
to follow	folgen
food	Nahrung, Essen
foot	Fuß
football	Fußball
for example	zum Beispiel
foreign	ausländisch, fremd
forever	für immer
to forget, (forgot, forgotten)	vergessen
fork	Gabel
to form; form	formen, gestalten; Form, Formular, Klasse
fortunate	glücklich
forward(s)	vorwärts
to free; free	befreien; frei
freeway	(AE) Autobahn
frequent	häufig
fresh	frisch
fridge	Kühlschrank
friend	Freund(in)
friendly	freundlich
front	Vorderseite
fruit	Frucht, Früchte, Obst
full	voll
fun	Spaß
furniture	Möbel
further	weiter
future	Zukunft; zukünftig
G	
game	Spiel
garage	Autowerkstatt, Garage
garden	Garten
gate	Tor
gentleman	Herr
to get, (got, got)	bekommen, werden
to get married	heiraten
to get up	aufstehen
girl	Mädchen
girlfriend	Freundin
to give, (gave, given)	geben
to give up	aufgeben
glad	froh
glass	Glas
glasses	Brille
glove	Handschuh
to go, (went, gone)	gehen
to go down	untergehen, sinken, fallen
to go on	weitergehen, weitermachen
to go shopping	einkaufen
to go up	hinaufgehen, wachsen, steigen
good	gut
goodbye	auf Wiedersehen
goods	Güter, Waren
to govern	regieren
government	Regierung
grandfather	Großvater
grandmother	Großmutter
grandparents	Großeltern
grateful	dankbar
great	großartig, groß
green	grün
to greet	grüßen
grey	grau
grocer	Lebensmittelhändler(in)
ground	Boden
group	Gruppe
to grow, (grew, grown)	wachsen
grown-up	Erwachsene(r); erwachsen
growth	Wachstum
to guess	erraten
guest	Gast
H	
hair	Haar
half	Hälfte; halb
hall	Flur, Saal
ham	Schinken
handbag	Handtasche
handkerchief	Taschentuch
to hang, (hung, hung)	(auf)hängen
to happen	geschehen
happy	glücklich, heiter
harbour	Hafen
hard	hart, schwierig, anstrengend
hardly	kaum
hat	Hut
to hate	hassen
to have, (had, had)	haben, lassen
to have to	müssen
to head; head	anführen, fahren nach; Kopf, Leiter
headache	Kopfschmerz
headline	Überschrift
health	Gesundheit
healthy	gesund
to hear, (heard, heard)	hören
heart	Herz
to heat; heat	heizen; Hitze
heating	Heizung
heavy	schwer
height	Höhe
hello	Hallo
to help; help	helfen; Hilfe
helpful	hilfreich
here	hier
here you are	bitte(schön)!
to hesitate	zögern
Hi Tech (High Technology)	Hochtechnologie
to hide, (hid, hidden)	(sich) verstecken
high	hoch
hill	Hügel
history	Geschichte
to hit, (hit, hit); hit	schlagen; Schlag, Treffer, Erfolg

to hold, (held, held)	halten
hole	Loch
holiday	freier Tag, Feiertag, Urlaub
home	Wohnung, Heimat; zu/nach Hause
homework	Hausaufgabe
honest	ehrlich
to hope; hope	hoffen; Hoffnung
horse	Pferd
hospital	Krankenhaus
hot	heiß
hour	Stunde
house	Haus
housewife	Hausfrau
housework	Hausarbeit
how	wie
how are you?	wie geht es Dir/Ihnen?
however	jedoch
hundred	Hundert
hungry	hungrig
to hurry; hurry	sich beeilen; Eile
to hurt	verletzen
husband	Ehemann
I	
ice cream	Eis
idea	Idee
if	wenn, falls
ill	krank
illness	Krankheit
to imagine	sich vorstellen
immediate	unmittelbar, unverzüglich
immigrant	Einwanderer, Einwanderin
to immigrate	einwandern
to import; import	importieren, einführen; Einfuhr, Import
importance	Wichtigkeit
important	wichtig
impossible	unmöglich
to improve	(sich) verbessern
in fact	tatsächlich
in front of	vor
in italics	kursiv
in my opinion	meiner Meinung nach
in time	rechtzeitig
inch	(Maßeinheit) Zoll
to include	einschließen
income	Einkommen
to increase	zunehmen, steigen, erhöhen
indeed	tatsächlich
industry	Industrie
to inform	informieren
inside	in, innerhalb
instead (of)	anstatt
to instruct	unterrichten, anweisen
instruction	Unterricht, Anweisung, (Gebrauchs-)Anleitung
insurance	Versicherung
interest	Interesse, (Plural) Zinsen
interesting	interessant
to introduce	einführen, vorstellen

introduction	Einführung, Vorstellung
to invent	erfinden
invention	Erfindung
to invest	investieren
invitation	Einladung
to invite	einladen
island	Insel
J	
jacket	Jacke
jam	Marmelade
job	Arbeit, Stelle, Aufgabe
to join	(sich) anschließen, beitreten
journey	Reise
juice	Saft
to jump; jump	springen; Sprung
just	genau, soeben, nur
K	
to keep, (kept, kept)	(be)halten
key	Schlüssel
to kill	töten
kilogram(me)	Kilogramm
kind	Art, Sorte; freundlich
king	König
to kiss; kiss	küssen; Kuss
kitchen	Küche
knee	Knie
knife	Messer
to knock; knock	klopfen; Klopfen
to know, (knew, known)	wissen
knowledge	Wissen
L	
ladder	Leiter
lady	Dame
lake	See
lamp	Lampe
land	Land
landlady	Vermieterin, Wirtin
landlord	Vermieter, Wirt
language	Sprache
large	groß
to last; last	dauern; letzte(r,s)
late	spät
later on	später
latest	neueste(r,s)
to laugh	lachen
laughter	Gelächter
law	Gesetz
to lay, (laid, laid)	legen
lazy	faul
to lead, (led, led)	führen, leiten
leaflet	(Hand-)Zettel, Flugblatt
to learn, (learnt, learnt)	lernen
learner	Lernende(r), Anfänger(in)
least	wenigste(r,s), geringste(r,s)

to leave, (left, left)	verlassen
left	links
leg	Bein
leisure	Freizeit
lemon	Zitrone
to lend, (lent, lent)	verleihen
length	Länge
less	weniger
lesson	Unterrichtsstunde
to let, (let, let)	lassen
letter	Brief
to lie, (lay, lain)	liegen
to lie, (lied, lied)	lügen
life	Leben
to lift; lift	anheben, aufheben; Aufzug, Mitfahrgelegenheit
light	Licht, Lampe; hell, leicht
to like; like	mögen; wie
limit	Grenze, Beschränkung
line	Linie, Zeile
lip	Lippe
to list; list	aufführen, auflisten; Liste
to listen	zuhören
little	klein, wenig
to live; live	leben, wohnen; lebend, direkt, live
living room	Wohnzimmer
to load; load	laden; Ladung
local	lokal, ortsansässig, örtlich
long	lang
to look	schauen, (aus-)sehen
to look after	sich kümmern um
to look for	suchen
to look forward to	sich freuen auf
to look out	aufpassen, hinaussehen
to look up	nachschlagen, heraussuchen, aufblicken
lorry	Lastwagen
to lose, (lost, lost)	verlieren
loss	Verlust
lot	Menge, Los
lots of	viel
loud	laut
to love; love	lieben; Liebe
lovely	hübsch
low	niedrig
luck	Glück, Schicksal
lucky	glücklich (im Sinne von: Glück haben)
luggage	Gepäck
lunch	Mittagessen
M	
machine	Maschine
mad	verrückt
madam	gnädige Frau
magazine	Zeitschrift
mail	Post
main	Haupt-
mainly	hauptsächlich
to make, (made, made)	machen, herstellen

man	Mann
to manage	zu Stande bringen, schaffen, leiten, verwalten
management	(Geschäfts-)Leitung, Verwaltung, Durchführung
many	viele
map	(Land-)Karte
market	Markt
marmalade	(Orangen-)Marmelade
married	verheiratet
to marry	heiraten
mass	Masse
to match; match	zusammenbringen, passen; Wettkampf, Streichholz
mathematics	Mathematik
to matter; matter	etwas ausmachen; Angelegenheit
may	dürfen
maybe	vielleicht
meal	Mahlzeit
to mean, (meant, meant)	bedeuten, meinen
meaning	Bedeutung
(in the) meantime	inzwischen
meanwhile	inzwischen
meat	Fleisch
mechanic	Mechaniker(in)
media	Medien
to meet, (met, met)	(sich) treffen, kennenlernen
meeting	Treffen
member	Mitglied
to mend	reparieren
to mention	erwähnen
message	Mitteilung, Nachricht
method	Methode
metre	Meter
microwave	Mikrowelle
middle	Mitte
midnight	Mitternacht
might	Gewalt, Macht; könnte(n)
mile	Meile
milk	Milch
to mind; mind	etwas ausmachen; Meinung, Gedanken, Verstand
mine	Bergwerk
minute	Minute
mirror	Spiegel
Miss	Fräulein
to miss	vermissen, verfehlen, verpassen
mistake	Fehler
to mix; mix	mischen; Mischung
moment	Augenblick
money	Geld
month	Monat
moon	Mond
more	mehr
morning	Morgen
most	die meisten
mother	Mutter
mountain	Berg
mouth	Mund
to move	bewegen, umziehen
movement	Bewegung

Mr	Herr
Mrs	Frau (verheiratet)
Ms	Frau (verheiratet oder unverheiratet)
much	viel
music	Musik
must	müssen
must not	nicht dürfen
N	
to name; name	benennen; Name
narrow	eng
near	nahe
nearby	nahe gelegen
nearly	fast, beinahe
necessary	notwendig
to need; need	benötigen; Notwendigkeit
neighbour	Nachbar(in)
neither … nor	weder … noch
never	nie
new	neu
news	Nachrichten, Neuigkeit
newspaper	Zeitung
next	nächste(r,s)
nice	nett
night	Nacht
no	kein; nein
no longer	nicht mehr
no one	niemand
nobody	niemand
noise	Lärm
noisy	geräuschvoll
noon	Mittag
north	Norden; nördlich, Nord-, nordwärts
nose	Nase
not	nicht
not either	auch nicht
not even	nicht einmal
not yet	noch nicht
to note; note	notieren; Notiz
nothing	nichts
to notice; notice	bemerken; Aushang, Kenntnis, Beachtung
now	jetzt
nowadays	heutzutage
nowhere	nirgendwo
number	Zahl, Nummer, Anzahl
nurse	Krankenschwester
O	
o'clock	Uhr (Zeitangabe)
ocean	Meer, Ozean
of course	natürlich
off	von … weg
to offer; offer	anbieten; Angebot
office	Büro, Amt
often	oft
oil	Öl
old	alt
on time	pünktlich

once	einmal
only	nur
to open; open	öffnen; offen
opinion	Meinung
opposite	Gegensatz; gegensätzlich
or	oder
to order; order	bestellen, befehlen; Bestellung, Befehl
organization	Organisation
to organize	organisieren
other	andere(r,s)
out of work	arbeitslos
outside	draußen, außerhalb; Außenseite
to own; own	besitzen; eigene(r,s)
owner	Eigentümer(in)
P	
to pack; pack	packen; Packung
packet	Paket
page	Seite
to paint; paint	malen; Farbe
pair	Paar
paper	Papier
parents	Eltern
to park; park	parken; Park
part	Teil
to pass	vorbeigehen, (Prüfung) bestehen
passenger	Passagier, Fahrgast
passport	Pass
past	Vergangenheit
to pay, (paid, paid); pay	(be-)zahlen; Bezahlung
peace	Friede
pen	Füllfederhalter
pence	Plural von Penny
pencil	Bleistift
penny	Penny (engl. Münze)
people	Leute, Volk
per cent	Prozent
perfect	vollkommen, perfekt
perhaps	vielleicht
period	Periode, Zeitraum, Schulstunde
personal	persönlich
petrol	Benzin
to phone; phone	telefonieren; Telefon
photograph	Foto
to pick	pflücken, wählen
to pick up	nehmen, aufheben
picture	Bild
piece	Stück
pipe	Pfeife, Röhre
to place; place	stellen, legen; Platz
to plan; plan	planen; Plan
plane	Flugzeug
to plant; plant	pflanzen; Pflanze
plate	Teller
platform	Bahnsteig
to play; play	spielen; (Schau-)Spiel
player	Spieler(in)
playground	Spielplatz
please	bitte

pleasure	Vergnügen
pocket	Tasche
to point; point	zeigen; Punkt
police	Polizei
policeman	Polizist
policewoman	Polizistin
political	politisch
politician	Politiker(in)
politics	Politik
to pollute	verschmutzen
polluted	verschmutzt
pollution	Umweltverschmutzung
poor	arm
popular	beliebt, populär
population	Bevölkerung
port	Hafen
possibility	Möglichkeit
possible	möglich
post office	Post
postcard	Postkarte
postman	Briefträger
pot	Topf
potato	Kartoffel
pound	Pfund
power	Kraft
power station	Kraftwerk
practice	Übung
to practise	üben
to prefer	vorziehen
to prepare	vorbereiten
to present; present	schenken, vorstellen; Geschenk, Gegenwart; anwesend, gegenwärtig
to press; press	drücken, pressen; Presse
pretty	hübsch
price	Preis
pride	Stolz
prize	Preis, Gewinn
probable	wahrscheinlich
to produce	produzieren, herstellen
producer	Hersteller(in), Produzent(in)
product	Produkt
production	Herstellung, Produktion
to profit; profit	profitieren; Gewinn
profitable	rentabel, einträglich, lohnend
program(me)	Programm
to promise; promise	versprechen; Versprechen
to pronounce	aussprechen
pronunciation	Aussprache
to protect	schützen
protection	Schutz
proud	stolz
pub	Kneipe
public	Öffentlichkeit; öffentlich
to pull	ziehen
pupil	Schüler(in)
purpose	Zweck, Absicht
to push; push	schieben, stoßen; Stoß
to put, (put, put)	setzen, stellen, legen
to put on	aufsetzen
to put up	(Hand) heben, bauen, aufstellen

Q	
quality	Qualität
quantity	Menge
quarter	Viertel
queen	Königin
question	Frage
quick	schnell
quiet	ruhig
quite	ganz, ziemlich
R	
railway	Eisenbahn
to rain; rain	regnen; Regen
to raise	(hoch-)heben, erheben
rather	ziemlich
to reach; reach	erreichen; Reichweite
to read, (read, read)	lesen
reader	Leser(in)
ready	fertig, bereit
real	wirklich
reality	Wirklichkeit
reason	Grund
reasonable	vernünftig, günstig
to receive	erhalten, bekommen
recent	jüngst, kürzlich
reception	Empfang
receptionist	im Empfang arbeitende Person
to record; record	aufnehmen; Rekord, Aufnahme
red	rot
regular	regelmäßig
relative	Verwandte(r); relativ
to relax	(sich) entspannen
to remain	bleiben
to remember	sich erinnern an
to remind (of)	erinnern an
to rent; rent	mieten, vermieten; Miete
to repair; repair	reparieren; Reparatur
to repeat; repeat	wiederholen; Wiederholung
to replace	ersetzen
to reply; reply	antworten; Antwort
to report; report	berichten; Bericht
to rescue; rescue	retten; Rettung
to rest; rest	ruhen; Rest, Ruhe
to result; result	resultieren; Ergebnis
to return; return	zurückkommen; Rückkehr, Rückgabe
rich	reich
to ride, (rode, ridden); ride	reiten, fahren; Ritt, Fahrt
right	rechts
to ring, (rang, rung); ring	klingeln, anrufen; Ring, Anruf
to rise, (rose, risen); rise	(auf)steigen, zunehmen; Aufstieg, Zunahme
river	Fluss
road	Straße
roof	Dach
room	Zimmer
round	Runde; rund, herum
rubber	(Radier-)Gummi

rubbish	Abfall, Müll
to rule; rule	herrschen; Regel, Herrschaft
to run, (ran, run)	laufen
S	
sad	traurig
safe	sicher
safety	Sicherheit
sale	(Schluss-)Verkauf
salesman/woman	Verkäufer(in)
salt	Salz
same	gleiche(r,s), der-, die-, dasselbe
to save	retten
to say, (said, said)	sagen
school	Schule
sea	Meer
seat	Sitz
second	Sekunde; zweite(r,s)
secret	Geheimnis; geheim
secretary	Sekretär(in)
to see, (saw, seen)	sehen
to seem	scheinen
to sell, (sold, sold)	verkaufen
seller	Verkäufer(in)
to send, (sent, sent)	schicken, senden
to sentence; sentence	verurteilen; Satz, Urteil
serious	ernst
to serve	(be)dienen, servieren
to service; service	(Auto) warten; Dienst, Betrieb
to set, (set, set)	setzen, stellen, legen
several	mehrere
shall	sollen
to share; share	teilen; Anteil
sheet	Blatt, Bogen
shelf	Regal
to shine, (shone, shone)	scheinen, leuchten
to ship; ship	verschicken; Schiff
shirt	Hemd
shoe	Schuh
shop	Geschäft, Laden
shop assistant	Verkäufer(in)
short	kurz
should	sollte(n)
shoulder	Schulter
to shout; shout	rufen; Ruf
to show, (showed, shown); show	zeigen; Ausstellung, Vorstellung
to shut, (shut, shut)	schließen
sick	krank
side	Seite
sight	Sicht, Anblick, Sehenswürdigkeit
to sign; sign	unterschreiben; Zeichen
simple	einfach
since	seit
since then	seitdem
to sing, (sang, sung)	singen
sir	Herr (Anrede)
sister	Schwester
to sit, (sat, sat)	sitzen
to sit down	sich hinsetzen
size	Größe

skill	Fertigkeit, Fähigkeit, Können
skirt	Rock
sky	Himmel
to sleep, (slept, slept); sleep	schlafen; Schlaf
slim	schlank
slow	langsam
to slow down	verlangsamen
small	klein
to smell; smell	riechen; Geruch
to smile; smile	lächeln; Lächeln
to smoke; smoke	rauchen; Rauch
to snow; snow	schneien; Schnee
so far	bisher, bis jetzt
soap	Seife
soccer	Fußball
sock	Socke
soft	weich
solution	Lösung
to solve	lösen
some	einige
somebody	jemand
someone	jemand
something	etwas
sometimes	manchmal
son	Sohn
song	Lied
soon	bald
sorry	betrübt, tut mir Leid!, Entschuldigung!
to sound; sound	klingen; Geräusch, Klang
south	Süden, südlich, Süd-, südwärts
space	Raum, Platz, Weltraum
spare time	Freizeit
to speak, (spoke, spoken)	sprechen
speaker	Sprecher(in)
special	besondere(r,s), spezielle(r,s)
speech	Rede
to speed; speed	schnell fahren; Geschwindigkeit
to speed up	beschleunigen
to spell; spell	buchstabieren; Zauber(spruch)
to spend, (spent, spent)	ausgeben, verbringen
spoon	Löffel
spring	Frühling
staff	Belegschaft, Kollegium
stamp	Briefmarke
to stand, (stood, stood); stand	stehen, aushalten; Stand
to start; start	anfangen; Anfang
to state; state	erklären, darlegen; Staat, Zustand
station	Bahnhof
to stay	bleiben
still	dennoch, noch
stone	Stein
to stop; stop	(an)halten; Halt
to store; store	lagern, speichern; (AE) Laden, Kaufhaus, Lager
storm	Sturm
story	Geschichte
strange	merkwürdig
street	Straße

to strike; strike	streiken, schlagen; Streik
strong	stark
to study; study	studieren; Studium
subject	Fach, Thema
suburb	Vorort
to succeed (in)	Erfolg haben (in, bei)
success	Erfolg
successful	erfolgreich
such	solche(r, s)
sudden	plötzliche(r, s)
sugar	Zucker
to suit; suit	passen; Anzug
suitcase	Koffer
sum	Summe
summer	Sommer
sun	Sonne
sunshine	Sonnenschein
supermarket	Supermarkt
supper	Abendessen
to suppose	annehmen, vermuten
sure	sicher
to surprise; surprise	überraschen; Überraschung
surprised	überrascht
survey	Überblick, Umfrage
sweet	süß
to swim, (swam, swum)	schwimmen
to switch; switch	schalten, wechseln; Schalter
T	
table	Tisch, Tabelle
to take, (took, taken)	nehmen
to take part in	teilnehmen an
to take place	stattfinden
to talk; talk	sprechen, reden; Gespräch
tall	hoch(gewachsen), groß
to taste; taste	schmecken, probieren; Geschmack
tasty	lecker
tax	Steuer
tea	Tee
to teach, (taught, taught)	lehren, unterrichten
teacher	Lehrer(in)
team	Mannschaft
technical	technisch
technology	Technik, Technologie
telephone	Telefon
television (TV)	Fernsehen
to tell, (told, told)	erzählen
terrible	schrecklich
to test; test	testen, überprüfen, untersuchen; Klassenarbeit, Versuch
than	als
to thank	danken
that	dass, jene(r, s), welche(r, s)
then	dann, damals
there	dort
there is/are	es gibt
therefore	deshalb
these	diese
these days	heutzutage
thick	dick

thin	dünn
thing	Ding
to think, (thought, thought)	denken
thirsty	durstig
this	diese(r, s)
those	jene
though	obwohl, trotzdem
thousand	Tausend
through	durch
to throw, (threw, thrown)	werfen
thus	so
ticket	Fahrkarte
tidy	ordentlich, aufgeräumt
till	bis
time	Zeit
times	Mal(e)
tired	müde
tobacco	Tabak
today	heute
toe	Zeh
together	zusammen
toilet	Toilette
tomato	Tomate
tomorrow	morgen
tongue	Zunge
tonight	heute Abend
too	auch
tool	Werkzeug
tooth, teeth	Zahn, Zähne
top	Spitze
tour	(Rund-)Reise, Tour
toward(s)	in Richtung
tower	Turm
town	Stadt
toy	Spielzeug
to trade (in); trade	handeln (mit); Gewerbe, Handwerk, Handel
traditional	traditionell, herkömmlich
traffic	Verkehr
traffic light(s)	Ampel
to train; train	ausbilden, trainieren, eine Ausbildung machen; Zug
training	Ausbildung, Training
to translate	übersetzen
translation	Übersetzung
to transport; transport	transportieren; Transport
to travel	reisen, fahren
tree	Baum
trip	(Kurz-)Reise
trouble	Schwierigkeit(en)
trousers	Hose
true	wahr, richtig
truth	Wahrheit
to try; try	versuchen; Versuch
to turn	(sich) drehen, wenden
to turn left/right	links/rechts abbiegen
to turn on/off	ein-/ausschalten
to turn over	umdrehen, überschlagen
twice	zweimal
to type; type	tippen, (Maschine) schreiben; Art, Typ

typewriter	Schreibmaschine
typical	typisch
typist	Schreibkraft
U	
umbrella	Schirm
uncle	Onkel
to understand, (understood, understood)	verstehen
unemployed	arbeitslos
unemployment	Arbeitslosigkeit
unfortunate	unglücklich
unfortunately	leider
unfriendly	unfreundlich
unhappy	unglücklich
university	Universität
until (till)	bis
upon	auf
upstairs	(nach) oben, im Obergeschoss
to use; use	gebrauchen; Gebrauch
used	gebraucht, gewohnt
useful	nützlich
useless	nutzlos
usual	gewöhnlich
V	
vacation	(AE) Ferien
van	Lieferwagen
vegetable(s)	Gemüse
vehicle	Fahrzeug
very	sehr
village	Dorf
to visit; visit	besuchen; Besuch
visitor	Besucher(in)
vocabulary	Wortschatz, Vokabelverzeichnis
voice	Stimme
to vote; vote	abstimmen, wählen; Abstimmung, Stimme
W	
to wait	warten
waiter	Kellner
waitress	Kellnerin
to wake, (woke, woken) (up)	wecken, aufwachen
to walk; walk	gehen; Spaziergang
wall	Wand, Mauer
to want	wollen
war	Krieg
warehouse	Lager
to warm; warm	wärmen; warm
to warn	warnen
to wash	waschen
washing machine	Waschmaschine
to waste; waste	verschwenden; Abfall, Verschwendung
to watch; watch	beobachten, sehen; Uhr
water	Wasser
way	Weg, Art und Weise
weak	schwach
to wear, (wore, worn)	(Kleidung) tragen
weather	Wetter
week	Woche
weekend	Wochenende
weight	Gewicht
to welcome; welcome	willkommen heißen; Willkommen, Empfang
well	gut, gesund; na ja …
well-known	bekannt
west	Westen, westlich, West-, westwärts
wet	nass, feucht
what	was
what about …	was ist mit …
what else	was noch
what for	wofür, wozu
wheel	Rad
when	wann
where	wo
whether	ob
which	welche(r,s)
while	während
white	weiß
who	wer
whole	ganz
whom	wem, wen
whose	wessen
why	warum
wide	weit
wife	Ehefrau
will	Wille; werden
to win, (won, won)	gewinnen
window	Fenster
wine	Wein
to wish; wish	wünschen; Wunsch
within	innerhalb
without	ohne
woman	Frau
wonderful	wunderbar
wood	Holz, Wald
wool	Wolle
word	Wort
to work; work	arbeiten; Arbeit
worker	Arbeiter(in)
world	Welt
worse	schlechter
worst	am schlechtesten
worth	wert
to write, (wrote, written)	schreiben
writer	Verfasser(in), Schriftsteller(in)
wrong	falsch
Y	
year	Jahr
yellow	gelb
yesterday	gestern
yet	jedoch, schon
you're welcome	gern geschehen!
young	jung
youth	Jugend

Irregular verbs

Infinitive	Past tense	Past participle	German
awake	awoke	awoken	aufwachen
be	was, were	been	sein
beat	beat	beaten	schlagen
become	became	become	werden
begin	began	begun	anfangen, beginnen
bend	bent	bent	biegen
bet	bet	bet	wetten
bid	bid	bid	bieten
bite	bit	bitten	beißen
blow	blew	blown	blasen, wehen (Wind)
break	broke	broken	brechen
bring	brought	brought	bringen
broadcast	broadcast	broadcast	senden (Rundfunk)
build	built	built	bauen
burn	burned/ burnt	burned/ burnt	(ver)brennen
burst	burst	burst	platzen
buy	bought	bought	kaufen
cast	cast	cast	werfen, gießen
catch	caught	caught	fangen
choose	chose	chosen	wählen, auswählen
come	came	come	kommen
cost	cost	cost	kosten
cut	cut	cut	schneiden
deal	dealt	dealt	handeln mit, umgehen
dig	dug	dug	graben
do	did	done	tun, machen
draw	drew	drawn	ziehen, zeichnen
dream	dreamed/ dreamt	dreamed/ dreamt	träumen
drive	drove	driven	fahren
drink	drank	drunk	trinken
eat	ate	eaten	essen
fall	fell	fallen	fallen
feed	fed	fed	füttern
feel	felt	felt	fühlen
fight	fought	fought	kämpfen
find	found	found	finden
fly	flew	flown	fliegen
forget	forgot	forgotten	vergessen
forgive	forgave	forgiven	vergeben
freeze	froze	frozen	gefrieren
get	got	got	bekommen
give	gave	given	geben
go	went	gone	gehen
grind	ground	ground	mahlen, schleifen
grow	grew	grown	wachsen
hang	hung	hung	hängen
have	had	had	haben
hear	heard	heard	hören
hide	hid	hidden	verstecken

Infinitive	Past tense	Past participle	German
hit	hit	hit	schlagen, treffen
hold	held	held	halten
hurt	hurt	hurt	verletzen
keep	kept	kept	behalten
know	knew	known	wissen, kennen
lay	laid	laid	legen
lead	led	led	führen
learn	learned/ learnt	learned/ learnt	lernen
leave	left	left	lassen, verlassen
lend	lent	lent	(ver)leihen
let	let	let	lassen
lie	lay	lain	liegen
lose	lost	lost	verlieren
make	made	made	machen
mean	meant	meant	bedeuten
meet	met	met	treffen
pay	paid	paid	bezahlen
put	put	put	setzen, stellen, legen
read	read	read	lesen
ride	rode	ridden	reiten, fahren
ring	rang	rung	klingen, anrufen
rise	rose	risen	aufgehen, steigen
run	ran	run	rennen, laufen
say	said	said	sagen
see	saw	seen	sehen
seek	sought	sought	suchen
sell	sold	sold	verkaufen
send	sent	sent	schicken, senden
shake	shook	shaken	schütteln
show	showed	showed/ shown	zeigen
shut	shut	shut	schließen
smell	smelt	smelt	riechen
sing	sang	sung	singen
sit	sat	sat	sitzen
sleep	slept	slept	schlafen
speak	spoke	spoken	sprechen
spend	spent	spent	ausgeben
stand	stood	stood	stehen
swim	swam	swum	schwimmen
take	took	taken	nehmen
teach	taught	taught	lehren, unterrichten
tear	tore	torn	zerreißen
tell	told	told	sagen, erzählen
think	thought	thought	denken
throw	threw	thrown	werfen
unterstand	understood	understood	verstehen
wake	woke	woken	wecken
wear	wore	worn	tragen (v. Kleidung)
win	won	won	gewinnen
write	wrote	written	schreiben

Unit 1

■ A2 Two students meet on the campus (see Student's Book, page 11)

■ D1 An exchange visit (Exercises a and b)

Mrs Evans: So what are you doing at College this week, John?
John: We're working on the environment project until Thursday, Mum. After that there's a visit to the electronics factory near here and to the power station in Hartlepool. I'm looking forward to it, I think it's really great that we have the chance to see places like that.
Klaus: Oh yes, I'm looking forward to it, too.
Mrs Evans: I'm glad you're enjoying this week, Klaus.
Klaus: Yes, I am. I'm learning such a lot about Britain and the way you do things here.
Mrs Evans: That's why these exchanges are such a good idea in my opinion. You young people have the chance to really get to know each other and that's important.
Klaus: Oh yes, I agree with you, Mrs Evans. I don't think that tourists learn half as much about the people or the country they visit.
John: Exactly. They don't have enough contact with the locals in my view. And when they come back they still think for example that all the Germans are serious and work hard.
Klaus: Or that the British are very traditional and drink tea all day!
Mrs Evans: But I suppose it's a bit strange for you, Klaus, living with a foreign family?
Klaus: No, not at all. I mean, you're all so friendly to me. Your house is so comfortable and I love the food. In fact, I feel quite at home already! Even the T.V. programmes are the same as the ones we have at home!

Unit 2

■ A2 A radio interview (see Student's Book, page 19)

■ D Job advertisements (Exercise d)

(Interview 1)
Interviewer: Well Miss Long, your CV seems alright. Tell me. What I'd like to know is which subjects did you enjoy most at school?
Miss Long: Ah … I liked foreign languages the most. We did French and German.
Interviewer: And can you speak them well?
Miss Long: Mm, my French isn't so good now, but I think my German is alright. I often spend my holidays in Germany as I have some friends over there and that gives me a chance to practise.
Interviewer: Hm. And why have you applied for this job?

Miss Long: Two main reasons really. First I'd like to travel in Europe more. And secondly, I'm only 22, and I feel it is time for a change. I'd like a more interesting job. As you know, I've worked as a secretary at Hard and Soft for 2 years now in the sales department. They produce machine parts for the car industry. I don't like the work ...

(Interview 2)

Interviewer: Mr Short. I see from your CV that you now work for a large electrical company. Why do you want to change jobs ?

Mr Short: Well, I'm 30 and I've worked at ElectroMagnum for 5 years. It's time for me to look for something else. I've been a computer operator for 8 years now.

Interviewer: Aren't you happy in your present job?

Mr Short: Well, not really. What I want is a new challenge. And the pay at ElectroMagnum isn't that good.

Interviewer: Well, we can talk about that later. Now, I see from your letter of application that you did a BTEC course in electrical engineering. What did you like most on that course?

Mr Short: The practical work was the really interesting part of the course, although I also enjoyed learning German. The trouble was I wasn't very good at it.

Interviewer: But you do know that this job involves travelling to Europe a lot and especially to our customers in Germany?

Mr Short: Yes, I know, and I am going to evening classes to improve my German. It's hard work though ...

UNIT 3

D1 Voting systems (Exercises a and c)

(Politician 1) Brian Taylor: "Good morning ladies and gentlemen. And welcome to Westminster. My name's Brian Taylor and before I became an MP, I ran my own farm in Scotland and farming is still one of my main interests. In a way, I suppose you could say that I was fortunate to get into Parliament in the first place. You see, I belong to the Liberal Democrats. We're one of the smallest parties in the British Parliament, not because we have the fewest voters, but because of the British voting system. In Great Britain the voters have one vote, which they can use for one particular candidate. The candidate who receives the most votes becomes the Member of Parliament for that constituency. I was lucky in 1992 when I won the election in Perth. But just look at these statistics ... "

(Politician 2) Inge Jensen: "Hello, I'm Inge Jensen from Copenhagen. Well, we in Denmark have a different voting system to that in Great Britain. Here, the party that wins most of the votes over the whole of Denmark has the most members in our Parliament – which is called the Folketing by the way. This is certainly more democratic because the smaller parties can get into Parliament more easily. Anyway, I expect you'd like to know more about me. Well, I joined the Danish Conservative Party while I was still at school and have been a member ever since. After I left school I got a job as a computer operator. However, politics remained one of my main interests and three years ago I became a member of the Danish Parliament. It has been hard work because I've got two children and a husband to look after as well as my work in Parliament. That's one reason why I'm especially interested in family affairs in politics. But let me go back to our voting system. Now, in my opinion, ..."

Unit 4

D1 Radio advertising (Exercises a and b)

1.

Keep on running, keep on hiding.
One fine day I'm gonna be the one to make you understand.
With Lukes you're gonna be a man!
Hey, hey, hey everyone is wearing them now. Nearly all the time. Hey, hey, hey your local store is stocking them now. At only £10.99.
Keep on running …

Lukes – the superfast running shoes. Really comfortable and built for speed! Get yours today at Simpson's Sports Store, Penarth Road.

2.

When you order your new car from Wyndham Jones, you can be confident that you're choosing from a world class range. You can be confident that every first-class car is totally safe and engineered to perfection. You can also be sure that at Wyndham Jones you will get excellent service. So whether you want a small or a family-sized car, at Wyndham Jones you know your choice will be the right one! Prices for the 'City' model start at £14,000. Wyndham Jones, Queen Street, Cardiff. Discover the difference!

3.

Shop assistant: Good afternoon sir, can I help you?

Customer: Oh, I'm looking for a personal CD player, actually. You know, one that you can carry around.

Shop assistant: I'm sure we can help you there, sir. Here at Sizzle superstore, we have a very large stock. This superb new Williams Mini-master model is very popular.

Customer: Yes, it's not bad, but it hasn't got many functions on it, has it? I mean, it's a bit basic, isn't it? Those Japanese ones, they're very good. Lots of functions on them. My mate at work bought one last week, he thinks it's marvellous.

Shop assistant: Alright sir, but I think you'll find that this Williams model is just as good. It's light and easy to use. It's got 20 functions and it only costs £99.

Customer: 99 pounds! Williams, you say, new company is it? Can't say I've heard of them. But 20 functions for 99 pounds, I'll take it!

Unit 5

D1 Travel announcements (Exercises a and b)

1.
The train now standing at platform 5 is the 10.20 to Liverpool, calling at Birmingham and Manchester. Passengers for Leeds and Newcastle please change at Birmingham.

2.
Flight BA 98 to Dublin is now ready for boarding. Would all passengers please proceed to Gate A6. We apologize for the delay which was due to fog at Dublin Airport. We hope you have a pleasant flight.

3.
Travel news! Trouble again on the M4 motorway near Bristol, where there has been a serious accident. Two lorries are now blocking the westbound section of the motorway and there is a three-mile traffic jam. The police advise motorists to leave the motorway at Exit 19 and to follow the diversion signs.

Unit 6

D1 Ecology or economics? (Exercises a and b)

Technologists, working for Zinc UK, a leading British mining company, have just published their plans for North Cornwall. They maintain that they have discovered large deposits of zinc ore in this beautiful part of the country not far from the seaside resorts of St. Agnes and Perranporth. The company believes that this is an extremely interesting commercial project because the ore contains a lot of zinc – up to 24 %. And according to Zinc UK there are about 50 million tonnes of the ore only 40 metres below the surface. Furthermore, Zinc UK emphasizes that this project will create about 200 new jobs in North Cornwall. However, not all the people in Cornwall are happy with the news. Local environmentalists and hotel owners have already expressed their fears. They are convinced that the mining project will destroy the area. The environmentalists believe that the beautiful countryside and the wild animals there are in danger. For this reason they have organized a demonstration to take place next week. Some hotel owners have joined the protest action because they fear that the mining company plans to turn the small seaside resorts into big ports for container ships to transport the zinc ore. According to hotel owners this will mean the death of the tourist resorts. So far Zinc UK has refused to make any further comment on this development.

Unit 7

D1 From poultry litter to electricity (Exercise a)

Chet: Now, this here's where our fuel arrives. The trucks drive straight in and drop the manure into those storage pits way down at the end there.
John: I see. And what about the people who live near here. Do they ever complain about the smell?
Chet: Well, no. We found a kind of neat solution to that problem. You see, the pressure here in the delivery hall is lower, so no fumes can get out. Now do you have your mask? We're going on through to the storage hall.
John: Do many people have to work in here?
Chet: No, there ain't many people here at all. From this point, the plant is fully-automated. Those conveyor belts there for example, which are taking the manure from the storage pits to burn in the furnace, are all computer-controlled.
John: And this furnace produces steam for the turbine, I suppose.
Chet: That's right. It gets real hot in that furnace – up to 800 degrees.
John: And what's your actual output here?
Chet: The generator produces 12.5 megawatts. That's enough electricity for 12,500 homes around here.
John: "Farmphos" – that's your fertilizer, isn't it? I could use some of that for my roses back home.

Chet: Yeah, I was fixing to give you a sack before you left. We produce it from the ash in the furnace. It's a quality product and very good for the environment …

Unit 8

D1 Sales talk (Exercises a and b)

Salesman: Good afternoon. Can I help you?
Customer: Well, actually I'm interested in one of your printers. You had an advert in the local paper.
Salesman: Ah, you mean the new laser printer. The Macprint XL7 .
Customer: Yes, that's right. Have you still got it in stock?
Salesman: Of course. If you'd like to come over here I can show it to you.
Customer: Hm, nice design.
Salesman: It not only looks good, it's one of the fastest printers on the market at the moment. It can do 20 pages per minute. And the print quality – I'll show you. What do you say to that?
Customer: Yes, it's pretty good.
Salesman: Pretty good? That's the best resolution you can get.
Customer: But what about paper? Can I use different types of paper?
Salesman: Of course. It even does envelopes as well. See this button here. You can decide what size paper you want to print, A4, A3, what you want.
Customer: That sounds good, but what about installing it? Can that be a problem?
Salesman: No, it's really easy. The handbook is so clearly written that even a complete beginner can understand it.

Customer: And what if something goes wrong with the printer?

Salesman: Well, that shouldn't happen. We at Microstore have a very good reputation, as you know. But if you did have a problem, we would come out right away and help you on the same day. You know, these big superstores are a little cheaper, but you don't get the same individual service as we offer.

Customer: Right. How much does one of these cost?

Salesman: It's yours for just £499 I think you'll agree that's a fantastic offer. A real bargain. And don't forget our guaranteed good service if anything goes wrong.

Customer: I'll have to think about it first.

Salesman: Yes, of course. May I give you my card? The name's John Clark. Please ask for me when you come in again.

Customer: Thank you very much.

Salesman: Thank you. May I give you this brochure …

Unit 9

D2 Looking for a new location (Exercises b and c)

Telephone call 1

Hello there. At the moment I'm here inside the busy shopping centre in Wetherby. Wetherby is a small town only 20 kilometres north west of Leeds, in the north of England. You've probably never heard of it but I can assure you Wetherby is a town which is developing quickly. And I think its success is mainly due to its excellent position. To give you a general idea – it's right next to the A1. And from that main road it's only half an hour's drive down to the M62 motorway which links the big ports of Liverpool and Hull. As I already said, Wetherby is expanding and the new industrial park just outside the town shows us what opportunities it in fact has. Most of the companies which have their sites there seem to be involved with light industry of one kind or another. The people in the development council I spoke to were very helpful and from my talks with them I found out that the price of land is much lower here than it is in the south of England. Despite the high unemployment rate in the area there's a positive atmosphere about the place. I would say, moreover, that the mixture of modern houses, good roads and clean modern industry makes Wetherby a pleasant place to live and work.

Telephone call 2

Hello there. Right now I'm sitting at the railway station here in Wolverhampton. Wolverhampton is about 20 kilometres north west of Birmingham, right in the middle of England.
Wolverhampton is a typical Black Country town, with both heavy and light industries, although most of the steelworks and coal mines have now been closed down. As a result Wolverhampton still has a rather high unemployment rate. The development council seems to be doing its best, but land still seems difficult to get and is also quite expensive. Most of the people I've spoken to live outside Wolverhampton. They said they prefer living in the new housing estates west of Wolverhampton rather than in the old terraced houses in town. One big advantage for companies here is that Wolverhampton is near the M6 motorway, which is a direct link to London and the Continent. This has attracted a lot of business to the area and has helped Wolverhampton to become a modern industrial centre. The visitor here gets the impression that Wolverhampton is a town which understands how to use the know-how and skills of its manufacturing tradition and at the same time be part of the 21st century.

Unit 10

D1 The new majority (Exercises a and d)

1.
I was told by the Russian authorities five or six times that I had no chance to get a visa. You see, back home in Minsk I was a painter. I was married and I had two sons. The only problem was – I was a Russian Jew. In spite of all the difficulties at that time I was reasonably happy, but it was my two sons who wanted to emigrate. In the end we got our visa and arrived here in San Francisco. With all our savings we opened a small fruit and vegetable store. But what I'm really proud of is how my two sons have come to terms with life in America. They are real Americans. I still have a few problems with the language but on the whole I feel happy here. Sometimes I get a little homesick, though.

2.
We came to America years ago from Cambodia as refugees. We had no luggage and no money. In Cambodia both my husband and I had worked in a restaurant until the Khmer Rouge army took over. We decided to leave our country and made our way to the refugee camps in Thailand. From there we got our visa for America. It was very hard for us at first, but now we've both got jobs near our apartment here in Los Angeles. I work on the production line of a chemical company. The pay is quite good – although it is a 45 hour working week. It's harder for my husband. He's got two part time jobs. In the mornings he works at a training center for Cambodians and the rest of the time he drives a truck for a food-processing company. He does not get home until 11 or 12 o'clock at night. That's not good for the family. The three children only really see him at weekends.

3.
I have been married now for 5 years but I've never met my wife's parents and she has not seen mine. You see, we arrived here from across the border seven years ago as so-called illegal immigrants and we do not dare leave America – we are afraid we will not be able to get across the border again. I have been working as a farm worker here in California and my wife has now got part-time work in a factory not far from our home in San Diego. Life is not easy as an "illegal" but what alternative do we have? You see, in Mexico you cannot find work – that is why we came here in the first place. And I think our daughter has a better future here than back home. Maybe we will get our visa one day.

Orkney
Islands
Outer
Hebrides
ATLANTIC
OCEAN
North
Sea
Inverness
Loch Ness
Aberdeen
Ben Nevis
Tay
Scotland
Glasgow
Clyde
Edinburgh
Londonderry
Northern
Ireland
Belfast
Carlisle
Tyne
Newcastle
Keswick
Lake
District
Cumbria
GREAT BRITAIN
Windermere
Kendal
Yorkshire
Isle of Man
Connemara
Irish Sea
Blackpool
Harrogate
York
Preston
Hull
REPUBLIC
OF IRELAND
Shannon
Dublin
Manchester
Leeds
Liverpool
Holyhead
Llanfair P.G.
Llandudno
Snowdon
Sheffield
England
Limerick
Nottingham
Leicester
Peterborough
Waterford
Rosslare
Dunmore East
Birmingham
Coventry
Wales
Solihull
Severn
Fishguard
Stratford-upon-Avon
Cambridge
Ipswich
Tredegar
Gloucester
Colchester
Harwich
Swansea
Oxford
Bladon
Headington
London
Thames
Cardiff
Bristol
Bath
Windsor
Gravesend
Somerset
Salisbury
Kent
Canterbury
Dover
Folkestone
Southampton
Exeter
Bournemouth
Portsmouth
Devon
Brighton
Eastbourne
Exmouth
Cowes
Weymouth
Isle of Wight
Cornwall
Torquay
Plymouth
English Channel
Land's End
Channel
Islands

U.S.S.R.
ARCTIC OCEAN
GREENLAND
ICELAND
Bering Str.
Queen Elizabeth Islands
Baffin Bay
Victoria I.
Baffin Island
Alaska
Yukon
Anchorage
Whitehorse
Juneau
Mackenzie
North West Territories
Rocky Mountains
Brit. Columbia
Alberta
Edmonton
Saskatchewan
Manitoba
Nelson
Hudson Bay
Labrador
Newfoundland
St. John's
Quebec
Ontario
Calgary
Regina
Winnipeg
Victoria
Seattle
Washington
New Brunswick
Fredericton
Halifax
Nova Scotia
Lake Superior
The Great Lakes
Ottawa
Maine
Montana
North Dakota
Minnesota
Oregon
Boise
Idaho
Yellowstone Nat. Park
Wyoming
South Dakota
The Great Plains
Wisconsin
Lake Michigan
Lake Huron
Michigan
Toronto
Lake Ontario
St. Lawrence R.
VT
NH
MA
Boston
NY
CT
RI
New York
East Orange
Detroit
L. Erie
PA
Pittsburgh
NJ
Philadelphia
DE
Washington D.C.
Sierra Nevada
Nevada
Utah
Colorado
Nebraska
Iowa
Mississippi
Missouri
Chicago
Illinois
Indiana
Ohio
Columbus
Dayton
WV
San Francisco
Fresno
California
Denver
Kansas
Indianapolis
St. Louis
KY
VA
Appalachian Mts.
ATLANTIC
Las Vegas
Grand Canyon
Mojave Desert
Painted Desert
Pawhuska
Nashville
NC
Los Angeles
Barstow
Flagstaff
Albuquerque
Amarillo
Oklahoma
Tulsa
Oklahoma City
TN
Oceanside
Phoenix
Arizona
New Mexico
Lawton
Arkansas
Birmingham
SC
Georgia
Alabama
Dallas
MS
Texas
OCEAN
LA
PACIFIC OCEAN
Lower California
Houston
New Orleans
Florida
Rio Grande
Miami
BAHAMAS
Gulf of Mexico
Havana
CUBA
MEXICO
Mexico City
0
1000 km
0
1000 Miles
CT Connecticut
DE Delaware
KY Kentucky
LA Louisiana
MA Massachusetts
MS Mississippi
NC North Carolina
NH New Hampshire
NJ New Jersey
NY New York
PA Pennsylvania
RI Rhode Island
SC South Carolina
TN Tennessee
VA Virginia
VT Vermont
WV West Virginia